Solar output and climate during the Holocene

Edited by Burkhard Frenzel

Co-edited by Teresa Nanni, Menotti Galli & Birgit Gläser

67 figures, 7 tables

SEMPER BONIS ARTIBUS

Gustav Fischer Verlag · Stuttgart · Jena · New York · 1995

Der vorliegende Sonderband wurde mit Mitteln der European Science Foundation (Straßburg) und der Akademie der Wissenschaften und der Literatur (Mainz) gefördert. Die Verantwortung für den Inhalt dieser Veröffentlichung liegt bei den Autoren.

Anschriften der Herausgeber:

Prof. Dr. Dr. h.c. Burkhard Frenzel, Institut für Botanik, Universität Hohenheim (210), D-70593 Stuttgart

Prof. Dr. Menotti Galli, Dipartimento di Fisica, Università di Bologna, Via Irnerio 46, I-40126 Bologna

Dr. Birgit Gläser, Institut für Botanik, Universität Hohenheim (210), D-70593 Stuttgart

Dr. Teresa Nanni, Istituto FISBAT-CNR, Via Gobetti 101, I-40126 Bologna

Sprachliche Beratung: Dr. Heike Neumann, Dept. of Geology, Cardiff University of Wales, GB
Bibliographische Bearbeitung: Dipl. Agr.-Biol. Mirjam Weiß, Hohenheim/ESF
Graphische Bearbeitung: Erika Rücker, Hohenheim
Technische Redaktion: Dr. Birgit Gläser, Hohenheim/ESF

Die Deutsche Bibliothek — CIP-Einheitsaufnahme

Solar output and climate during the Holocene : 7 tables / ed. by
Burkhard Frenzel. Co-ed. by Teresa Nanni ... — Stuttgart ; Jena ;
New York : G. Fischer, 1995
    (Palaeoclimate research ; Vol. 16 : ESF project "European
    palaeoclimate and man" ; Special issue 11)
    ISBN 3-437-30815-7
NE: Frenzel, Burkhard [Hrsg.]: Paläoklimaforschung / ESF project
    "European palaeoclimate and man"

ISBN 3-437-30815-7
US-ISBN 1-56081-443-0
ISSN 0930-4673

Paläoklimaforschung · Palaeoclimate Research

Akademie der Wissenschaften und der Literatur

Paläoklimaforschung
Palaeoclimate Research   Volume 16

Special Issue: ESF Project
"European Palaeoclimate and Man" 11

Editor: Burkhard Frenzel

Associate Editor: Birgit Gläser

1995

European Science Foundation
Strasbourg

Akademie der Wissenschaften
und der Literatur · Mainz

CONTENTS

Addresses of the authors

Prof. Dr. G. Bonino, Istituto di Cosmogeofisica del CNR, Corso Fiume 4, I-10133 Torino

Dr. D. Camuffo, CNR-ICTIMA, Corso Stati Uniti, 4, I-35020 Padova

Dr. S. Cecchini, Istituto TESRE-CNR, Via Gobetti 101, I-40126 Bologna

Prof. Dr. G. Cini Castagnoli, Istituto di Cosmogeofisica del CNR, Corso Fiume 4, I-10133 Torino

Dr. H. B. Clausen, The Niels Bohr Institute of Physics, Astronomy and Geophysics, Department of Geophysics, Univ. of Copenhagen, Haraldsgade 6, DK-2200 Copenhagen

Dr. W. Dansgaard, The Niels Bohr Institute of Physics, Astronomy and Geophysics, Department of Geophysics, Univ. of Copenhagen, Haraldsgade 6, DK-2200 Copenhagen

Dr. S. Enzi, CNR-ICTIMA, Corso Stati Uniti, 4, I-35020 Padova

Prof. Dr. M. Eronen, Department of Geology, Univ. of Oulu, Linnanmaa, FIN-90570 Oulu

Dr. T. Fichefet, Université Catholique de Louvain, Institut d'Astronomie et de Géophysique G. Lemaître, chemin du Cyclotron 2, B-1348 Louvain-la-Neuve

Prof. Dr. C. Fröhlich, Physikalisch-Meteorologisches Observatorium Davos, World Radiation Center, CH-7260 Davos Dorf

Prof. Dr. M. Galli, Dipartimento di Fisica, Università di Bologna, Via Irnerio 46, I-40126 Bologna

Dr. N. Gundestrup, The Niels Bohr Institute of Physics, Astronomy and Geophysics, Department of Geophysics, Univ. of Copenhagen, Haraldsgade 6, DK-2200 Copenhagen

Dr. C. U. Hammer, The Niels Bohr Institute of Physics, Astronomy and Geophysics, Department of Geophysics, Univ. of Copenhagen, Haraldsgade 6, DK-2200 Copenhagen

Prof. Dr. M. K. Hughes, Laboratory of Tree-Ring Research, Univ. of Arizona, USA-Tucson, AZ 85721 (address for mail), and Cooperative Institute for Research in Environmental Science, Univ. of Colorado, USA-Boulder, CO 80309

Dr. S. J. Johnsen, The Niels Bohr Institute of Physics, Astronomy and Geophysics, Department of Geophysics, Univ. of Copenhagen, Haraldsgade 6, DK-2200 Copenhagen

Dr. J. Lipp, Institut für Hydrologie, GSF-Forschungszentrum für Umwelt und Gesundheit, Ingolstädter Landstr. 1, D-85764 Oberschleißheim

Dr. T. Nanni, Istituto FISBAT-CNR, Via Gobetti 101, I-40126 Bologna

Dr. E. Nesme-Ribes, URA 326, Observatoire de Paris, 5 Place Janssen, F-92195 Meudon

Dr. A. Provenzale, Istituto di Cosmogeofisica del CNR, Corso Fiume 4, I-10133 Torino

Prof. J. G. Roederer, Geophysical Institute, University of Alaska Fairbanks, Fairbanks AK 99775-7320, USA

Dr. R. Sadourny, Laboratoire de Méteorologie Dynamique, CNRS, Ecole Polytechnique, F-91128 Palaiseau Cedex

Dr. M. Stuiver, Department of Geological Sciences and Quaternary Research Center, Univ. of Washington AK-60, USA-Seattle, WA 98195

Dr. H. Tauber, The Niels Bohr Institute of Physics, Astronomy and Geophysics, Department of Geophysics, Univ. of Copenhagen, Haraldsgade 6, DK-2200 Copenhagen

Dipl.-Phys. P. Trimborn, Institut für Hydrologie, GSF-Forschungszentrum für Umwelt und Gesundheit, Ingolstädter Landstr. 1, D-85764 Oberschleißheim

Prof. Dr. C. Vita-Finzi, Univ. College London, Department of Geological Sciences, Gower Street, GB-London WC1E 6BT

Dr. P. Zetterberg, Karelian Institute, Univ. of Joensuu, P.O. Box 111, FIN-80101 Joensuu

Preface

Teresa Nanni, Menotti Galli & Burkhard Frenzel

The scientific programme "European palaeoclimate and man since the last glaciation" (EPC) was launched in 1989. The main goal of the programme is to obtain a better understanding of the European climate variation in the last 11,500 years. The causes of climate variation during this epoch are of anthropogenic as well as of natural origin. The anthropogenic influence can be traced by the stages of human civilization especially from early Neolithic times. However, there is no information regarding the external forcing factors acting on our climate such as solar variability, volcanic or tectonic activity and, on a longer time-scale, orbital changes and celestial body impacts.

We are aware that solar output in the form of thermal and non-thermal radiation and particle emission is by far the most important forcing factor. The climate system, however, depending on the solid, liquid and gaseous phases and their spatial structures, is so complex that so far only a few climatic phenomena have been directly and convincingly related to solar activity.

"It is only in recent years that the availability of statistically robust results and the formulation of plausible trigger mechanisms to explain the observed correlation have placed the field in a more respectable standing with the general scientific community." As a consequence "the study of solar variability effects on the immediate human environment is now an important and integral part of the ICSU/SCOSTEP Solar Terrestrial Energy Programme 1990-1997 (STEP)" (ROEDERER, this volume).

The papers collected in this volume present and discuss solar data and climatic proxy data of the past which can be used to search for solar activity signatures and some aspects of the sun-earth physics. After a review on the state of the advancement of the problem by ROEDERER, reconstructed climatic data of the past, i.e. temperatures, are reported. They are deduced from $\delta^{18}O$ values in the Greenland ice core experiment GRIP (DANSGAARD, JOHNSEN, CLAUSEN, GUNDESTRUP, HAMMER & TAUBER), δD measured in tree-rings (LIPP & TRIMBORN), tree-ring width or maximum density chronologies (HUGHES; ERONEN & ZETTERBERG). A discussion about temperatures and other phenomena during the cold season and their possible connection to solar variability is performed by CAMUFFO & ENZI, VITA-FINZI, and CECCHINI, GALLI & NANNI. The solar signal in the Mediterranean sea sediments (CINI-CASTAGNOLI, BONINO & PROVENZALE) and in long tree-ring ^{14}C series (STUIVER) is also investigated as radiocarbon variation is related to modulation effects of solar output variation. An important discussion is presented by FRÖHLICH on how to extrapolate the satellite measurements of total solar irradiance and its variability back to Maunder Minimum times, according to recent views on solar physics. A possible connection between the Maunder Minimum of solar activity and the deepest phase of the Little Ice Age is discussed by NESME-RIBES and SADOURNY. Finally, results of a model computation of the global climatic effect due to a change of total solar irradiance are reported by FICHEFET.

A fairly convincing demonstration of a direct solar influence on climate that will allow to make predictions is not yet available. However, many encouraging hints pointing to future developments in this field have been obtained.

Solar variability effects on climate

Juan G. Roederer

Summary

One of the key tasks in climate research at the present time is to separate anthropogenic effects from natural change. Solar variability is one possible cause of natural change in addition to other external phenomena such as volcanic eruptions and, on a long-term scale, orbital changes and celestial body impacts. The topic of "sun-weather relationships" has followed a long and tortuous history of scientific speculation and controversy since the continuous observation of sunspots began in the seventeenth century. It was not until recently that more systematic studies with long-term data bases of meteorological, climatological and solar parameters led to an increasing, statistically robust, body of evidence for a causal connection between some manifestations of solar variability and changes in the troposphere and the climate system. While of planetary scale, the strength and sign of pertinent correlations have distinct geographic, seasonal and other temporal characteristics. Most likely, several trigger mechanisms are at work simultaneously, but their relative importance may depend on the time scale envisaged and on competing processes such as volcanic eruptions. In this paper we will summarize the most frequently formulated criticisms, review the most recent results on relevant solar variability effects at the 11-year, secular, and short-term time scales, and discuss proposed mechanisms such as enhanced Hadley circulation, charged particle ionization effects on clear-air conductivity, and electric field effects on the microphysics of cloud formation.

Zusammenfassung

Die Trennung anthropogener Einflüsse von natürlichen Klimaveränderungen ist gegenwärtig eine der Hauptaufgaben in der Klimaforschung. Neben externen Phänomenen wie Vulkanausbrüchen und - über längere Zeitabschnitte betrachtet - Veränderungen in der Erdumlaufbahn sowie Meteoriteneinschlägen sind Änderungen der Sonnenaktivität eine mögliche Ursache natürlicher Klimaschwankungen. Seit im siebzehnten Jahrhundert die kontinuierliche Beobachtung der Sonnenflecken begann, entwickelte sich die Frage nach der Beziehung von Sonne und Klima zu einem Objekt wissenschaftlicher Spekulation und Kontroverse, das eine lange und komplizierte Geschichte aufweist. Erst kürzlich führten systematische Untersuchungen, die auf langzeitig dokumentierten meteorologischen, klimatologischen und solaren Parametern basieren, zu wachsendem, statistisch abgesichertem Beweismaterial für eine kausale Beziehung zwischen bestimmten Variationen der Sonnenaktivität und Veränderungen in der Troposphäre und dem Klimasystem. Obgleich sie sich im planetarischen Maßstab abspielen, haben die betreffenden Korrelationen klare geographische, jahreszeitliche oder andere zeitliche Charakteristika. Höchstwahrscheinlich sind mehrere auslösende Mechanismen gleichzeitig wirksam, ihre jeweilige Bedeutung hängt jedoch von der zu betrachtenden Zeitspanne und von möglicherweise gleichzeitig ablaufenden Prozessen wie z.B. Vulkanausbrüchen ab. In diesem Beitrag werden die am häufigsten formulierten Kritikpunkte zusammengefaßt, die jüngsten Ergebnisse über die Auswirkungen der solaren Variabilität in 11-jährigen, hundertjährigen und kurzfristigen Zeitintervallen dargestellt und bestimmte Mechanismen diskutiert, wie beispielsweise eine verstärkte Hadley-Zirkulation, die Auswirkungen geladener energetischer Teilchen auf die Leitfähigkeit wolkenfreier Luft und die Auswirkungen des elektrischen Feldes auf die mikrophysikalischen Prozesse der Wolkenbildung.

1. Introduction

The topic of "sun-weather relationships" has followed a long and tortuous history of scientific speculation and controversy. As soon as sunspots became the subject of systematic observation after Galileo's invention of the telescope, scientists began to wonder about the possible effects of these "blemishes" on terrestrial weather. Yet from the very beginning, the efforts to discover sun-weather relationships carried the stigma of "bad science". This was in part due to the fact that the only scientific paradigm available was to correlate selected meteorological parameters with time-dependent features on the sun - and hope that the correlation would hold up in the future. And in most cases it didn't!

It was not until the first decades of this century that more systematic studies were begun using long-term data bases of meteorological, climatological and solar parameters. An early example is the study of the global distribution of annual rainfall difference between sunspot cycle maximum and minimum, given in the book by CLAYTON (1923); later, ROBERTS (1975) discussed statistics on the occurrence of droughts in Nebraska, which showed a tendency to occur near that sunspot minimum which followed a magnetically negative maximum (of the 22-year Hale cycle). CLAYTON's study clearly showed that if a physical relationship existed between the 11-year sunspot cycle and tropospheric phenomena, it had to exhibit a regional dependence. ROBERTS' paper brought in the overall magnetic configuration of the sun (because of the correlation with the Hale cycle), and it also showed a phase-locking to the quasi-periodic character of the solar cycle (frequency modulation), a statistically significant fact given the low probability that an atmospheric periodicity unrelated to the sun would track the changes in solar cycle length just by chance.

During the seventies it became apparent that it was necessary to clearly divide the study according to three classes of solar variations: (1) the 11-year sunspot cycle (and, eventually, the related 22-year magnetic Hale cycle); (2) the long-term secular changes of sunspot cycle amplitude (the Gleissberg "cycle"); (3) short-duration, sporadic events, such as solar flares and the sun-controlled solar wind shocks and reversals of the interplanetary magnetic field. Concerning secular effects, EDDY (1976) published a classical paper revealing the remarkable correlation of winter temperatures in London and Paris with the Gleissberg cycle (traced back in time to the twelfth century by using as proxy indicator of solar activity the ^{14}C concentration in tree-rings). This paper received particular attention because it linked the "Little Ice Age" in the seventeenth century and the cold period in the fifteenth century with the Maunder and Spörer Minima, respectively, in which few sunspots were seen for several decades. Regarding short-term variability, WILCOX et al. (1974) published a study of changes of the northern hemisphere "vorticity area index" ("VAI", a quantitative measure of low-pressure troughs) at the times of magnetic sector boundary passages (interplanetary magnetic field reversals), showing statistically significant decreases of the VAI from two days before to one day after the passage during winter months (this correlation, however, did not subsist during the eighties).

These and many other studies (e.g., HINES & HALEVY, 1977; BUCHA, 1988) could not dispel the general scepticism about the subject *per se*, particularly on part of the meteorological community. It is only in recent years that the availability of statistically robust results and the formulation of plausible trigger mechanisms to explain the observed correlations have placed the field in a more respectable standing with the general scientific community.

The study of solar variability effects on the immediate human environment is now an important and integral part of the ICSU/SCOSTEP Solar Terrestrial Energy Program 1990-1997 (STEP) (ROEDERER, 1992). The present paper will discuss the most frequently formulated criticisms of solar activity-climate studies and review some important recent results. For detailed literature references, the reader is referred to the articles cited in this paper.

2. Most frequent criticisms and "freak facts of nature"

The study of solar variability effects on climate has been subjected to severe criticism that can be grouped into several distinct categories. While this criticism was well founded on historical grounds, much of it can now be dispelled on the basis of recent studies.

(1) *The power involved in solar variability is too small!* Indeed, the total solar electromagnetic power deposited in the earth's environment is more than 10^{12} MW, whereas the variable components of electromagnetic and particle energy impinging on the earth system represent only 10^4 to 10^6 MW, i.e., only a tiny fraction of the main energy flux. Furthermore, the variable energy components mainly affect the magnetosphere and upper atmosphere which are only very weakly coupled to the troposphere. The energy argument, however, is not valid for highly non-linear, complex systems such as the coupled atmosphere-ocean-biosphere. It is well known that complex systems can behave

chaotically, i.e., follow very different paths after the smallest change in initial or boundary conditions, or in response to the smallest perturbation. In a highly non-linear system with large reservoirs of latent energy such as the atmosphere-ocean-biosphere, global redistributions of energy can be triggered by very small energy inputs, a process that depends far more on their spatial and temporal pattern than on their magnitude. Moreover, such a system may exhibit sudden transitions between some "eigenstates" of quasi-equilibrium; recent data from the Greenland ice sheet clearly show this for the northern hemisphere atmosphere prior to the Holocene period (DANSGAARD et al., 1993).

(2) *We cannot think of any mechanisms responsible for solar-climate effects!* This is an "unscientific" argument; there are abundant historical examples of processes which were studied, accepted or used, but for which the responsible mechanism was not known for a long time (e.g., dinosaur extinctions, plate tectonics, or atomic spectral line emissions which, although "prohibited" by classical electrodynamics, had been in use in science and technology many decades before their explanation by quantum mechanics). Besides, the argument as a whole does not hold anymore: trigger mechanisms are presently being formulated and studied.

(3) *It's all just a coincidence!* Historically a quite valid argument, but now there are too many statistically robust results the probability of which to occur all by chance is extraordinarily small. This in particular applies to those correlations which stay in phase with the quasi-periodicity of solar activity. Predictions based on recent studies covering a limited period of time were verified when new, later, data came in (e.g., LABITZKE & VAN LOON, 1993) or when more data reaching into the past were included (e.g., FRIIS-CHRISTENSEN & LASSEN, 1993).

(4) *The intervals for which data are available for correlation studies are too short!* This is a valid criticism when applied to some of the recent studies of 11-year cycle effects. The band-width of uncertainty in the determination of the period using data from a time-interval of, say, only three solar cycles is ± 2 years; therefore, any intrinsic atmospheric periodicity between 9 and 13 years could be expected to show an acceptable correlation with the solar cycle (e.g., see the discussion in SALBY & SHEA, 1991). Until now, however, no plausible decadal cyclic variation of the atmosphere unrelated to the sun has been identified or proposed.

(5) *The "So what?" question.* It has been argued that whereas solar variability-climate effects do exist, they are so small that they are unimportant for forecasting purposes (e.g., PITTOCK, 1979). Leaving aside the question of whether they are really that small, this is yet another "unscientific" argument: every global and coherent response should be considered important for the physical understanding of the atmospheric system!

In addition to the above criticisms which the researchers in this field may continue to face, there are some "freak facts" of Mother Nature that conspire to make physical interpretations especially difficult.

(1) All solar emissions exhibit 11-year periodicity; this makes source identification more difficult. For instance, could an 11-year periodicity in the atmosphere be due to the 11-year variability of solar irradiance, or could it be due to the cumulative effect of short-term events (like solar flares), whose occurrence also exhibits an 11-year periodicity?

(2) The amplitudes of the last two solar maxima (both in terms of sunspot number and total irradiance) were nearly identical. This makes it difficult to model solar irradiance (for which absolute satellite measurements are available only after 1978) and its relation to sunspot number, a fact that in turn prevents using the latter as a proxy to estimate total solar irradiance values in the past.

(3) The time lag between annual average temperature over land and the mean sea-surface temperature (governed by the ocean response time), and the time lag between the 11-year running means of solar cycle amplitude (the Gleissberg cycle) and solar cycle duration, are both of the order of 10-15 years. This obscures the discriminability among proposed mechanisms (see Section 4.2).

(4) The beat period between the 26-28 month quasi-biennial oscillation (QBO) and the biennial period (24 months) is 12-14 years, i.e., of the same order as the solar cycle period. In principle, this can decrease statistical discriminability in QBO-stratified time-series (aliasing effect, TEITELBAUM & BAUER, 1990), but it is of little importance due to the strongly quasi-periodic character of the QBO (see Section 4.1).

(5) Global circulation model (GCM) calculations show that random stochastic variations of the mean atmospheric temperature can have similar amplitudes and time scales as the observed variations. This makes the distinction between unpredictable intrinsic varia-tions and those driven by external and anthropogenic forcing difficult.

(6) The "ultimate freak fact" is that key parameters such as the global temperature, sunspot cycle amplitude, greenhouse gas concentrations and sunspot cycle frequency all show a net increase since the late 1800s. This poses extra difficulties in the determination of the differential climate sensitivity to the various possible forcing factors.

3. The variable input channels

Figure 1 depicts schematically the most important energy input channels from the sun, their continuous or sporadic variability and the regions in the terrestrial atmosphere that can be affected by this variability. Solar magnetohydrodynamics regulates all forms of variability, including that of the intensity of the cosmic ray flux, the configuration of the high-latitude ionospheric electric field and the precipitation of energetic trapped electrons, all of which are controlled by the interplanetary magnetic field (IMF). The variability of the two photon channels at left is controlled by surface structures, such as faculae and plages (radiation emitters) and sunspots (blockers), whereas the particle channels are mainly controlled by processes in the coronal plasma. Finally, the earth's internal magnetic field, which is vari-able on a secular scale (including an approximately 6000-year oscillation of the main dipole moment), also influences the fluxes of the charged particles penetrating the atmosphere. Not considered at all in the figure (and in this paper) are the variations of solar insolation due to periodic changes in orbital and rotational parameters of the earth (leading to the Milankovitch cycles). Also not considered are the natural 27-day periodicities related to solar rotation. The atmospheric processes sketched in Fig. 1 are discussed in Chapter 5.

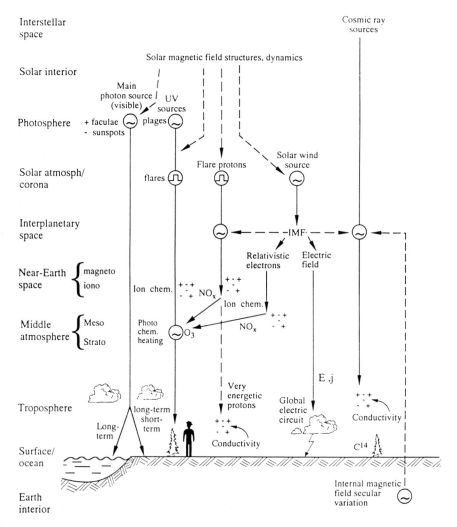

Fig. 1 Sketch of the channels of energy input from the sun (and beyond), indicating their continuous or sporadic variability (wavy and square-wave signs respectively), and the regions in the terrestrial atmosphere that can be affected by this variability.

An important discovery (e.g., Willson & Hudson, 1988) was the small but significant 11-year modulation of the "solar constant" (total irradiance in W/m^2 at earth). Figure 2 (from

KYLE et al., 1993) shows monthly mean values of the total irradiance as measured with a radiometer on NIMBUS 7, together with the Wolf sunspot number. The 11-year modulation comes from two sources of mutually counter-acting effects: enhanced emissions from bright faculae during solar maximum and enhanced blocking by sunspots (obviously the former wins over the latter). In addition, a long-term variation is speculated to exist, caused by the possible growth and decay of the global facular network on the solar surface (WHITE et al., 1992). Figure 2 shows that the overall relationship between total irradiance and sunspot number during one cycle is clearly non-linear and is different during the descending and ascending phases of solar activity. It is important to note that the UV band of the photon spectrum exhibits a much larger 11-year variability than that shown for the total irradiance (DONNELLY, 1991).

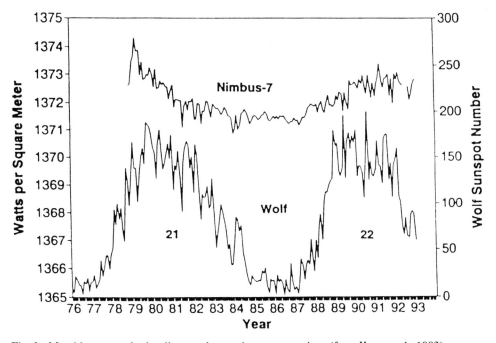

Fig. 2 Monthly mean solar irradiance values and sunspot numbers (from KYLE et al., 1993)

Since the two recent solar maxima shown in Fig. 2 are of nearly the same amplitude ("freak fact N° 2"), nothing can be concluded from these measurements about the long-term variation of irradiance. Using the satellite measurements and an isolated value of solar irradiance obtained by KOSTERS & MURCRAY (1979) in a 1968 balloon flight, REID (1991) established a linear relationship between total solar irradiance and the envelope of the 11-year sunspot cycle (which he used in model calculations of long-term change of sea surface temperatures; see Section 4.2). This relationship, however, would imply a value of the solar constant 11 W/m^2 lower during the cold period of the 1600s, an unrealistically large change. LEAN et al. (1992), for instance, estimate that the excess facular radiation from a complex magnetic surface configuration as it exists in the contemporary sun contributes about 1.5 W/m^2. However, using observations from solar-like stars, these authors conclude that for non-cycling stars (e.g., in a Maunder Minimum-like state) the irradiance should be further reduced, below that derived for a total removal of network magnetic flux; their final estimate of the Maunder Minimum total irradiance reduction is about 2.7 W/m^2 below the contemporary solar minimum value. These estimations are quite significant from the climatic point of view: GCM calculations show that at least half of the global temperature increase since the Little Ice Age could be explained by solar radiative forcing (RIND & OVERPECK, in print).

Coronal holes, coronal mass ejections and solar flares are transient manifestations of solar activity at different time scales; although in themselves aperiodic, they do occur with frequencies tied to the 11-year cycle. They all have important effects on solar wind density, speed and magnetic field configuration (the IMF). Energetic protons from large flares can penetrate the atmosphere down to sea level; shock waves emitted by flares "sweep away" cosmic rays, causing important decreases of their intensity at earth (the so-called Forbush decreases); sudden changes in the direction of the IMF alter the transfer of solar wind energy into the magnetosphere and can cause magnetic storms and aurorae; changes in the IMF also alter the electrical connection between the solar wind and the polar cap ionosphere, thus affecting the "global electric circuit" in the entire atmospheric system. The higher solar wind speed and the enhanced magnetic irregularities during solar maximum are responsible for a general decrease of the cosmic ray flux, which therefore is anti-correlated to solar activity. During magnetic storms, the magnetosphere "squeezes out" high energy electrons transiently trapped in the earth's magnetic field (BAKER et al., 1987), which subsequently precipitate into the upper atmosphere, preferentially in the region of the so-called South Atlantic Anomaly (where the Van Allen radiation belt comes closest to the atmosphere).

Figure 3 shows the "good old" monthly average sunspot number curve. This is the only indicator of solar activity that can be traced back reliably to the seventeenth century (NESME-RIBES, 1994). It is important to note, however, that it does not represent equally well all aspects of solar variability: the solar activity at different sunspot minima and its direct effects such as geomagnetic disturbance may be rather different, even if it does not manifest itself in the observed sunspot number directly. The sunspot number is not a single-frequency sinusoidal function of time; rather it has a notable amplitude modulation (given by the envelope of the curve in Fig. 3) and a frequency modulation as well (solar cycle length values ranging from 9-12 years). Notice also that the ascending and descending phases have varying slopes. Thus, solar cycle amplitude, length and, for instance, maximum time rate of change are important parameters for solar-terrestrial correlation studies. The envelope of the sunspot curve of Fig. 3 follows by 10-20 years a similar-looking curve of (minus) the solar cycle length (shown in Fig. 8); this is approximately the same time-lag that exists between the average global land and sea-surface temperatures ("freak fact N° 3"). Today, the 10.7 cm radiowave flux is often used instead of sunspot number as a measure of solar activity.

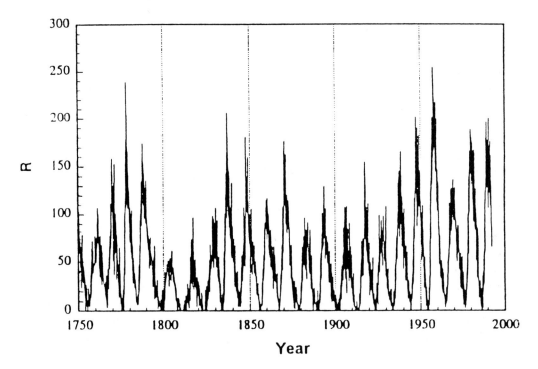

Fig. 3   The monthly mean sunspot number R during 2 1/2 centuries

4. Recent results

4.1 11-year cycle

A breakthrough in the study of solar variability effects on climate came in 1987. K. Labitzke of the Free University of Berlin had been engaged in a systematic study of the northern polar stratosphere and its relationship to the quasi-biennial oscillation (QBO) of the equatorial stratosphere (LABITZKE, 1982). The QBO refers to the winds at different layers of the tropical stratosphere, which reverse their direction (east or west) with a quasi-periodicity of about 26-28 months (the higher layers leading the lower ones). For the winter hemisphere, this really represents two possible dynamic states of the stratosphere, one in which the equatorial part of a given layer rotates in the same sense as the polar vortex (the west QBO phase) and one in which it contra-rotates; each state presents very different conditions for the propagation of kinetic and thermal energy and momentum toward the higher latitudes. In particular, it was found that the polar vortex was strong and stable, thus colder, during the corotating west phase than during the contra-rotating east phase when major disruptions (major mid-winter warmings) tended to occur in the polar stratosphere. But there were winters in which the reverse situation arose; they happened to correspond to epochs of solar maximum. Indeed, LABITZKE (1982) found that all major mid-winter warmings that occurred during the west phase happened only at times of solar maximum.

Analyzing the temperature of the 30 mb level at the North Pole during January and February for three solar cycles, LABITZKE (1987) found an astounding correlation/anti-correlation with solar activity (as expressed by the 10.7 cm flux) when the data were stratified according to the west/east phase of the QBO (defined by the equatorial wind at the 50-40 mb level). Figure 4 is an updated version of these results. It is thus clear that high solar activity introduces a radical perturbation in the process responsible for the global coupling between the tropical and the polar regions of the stratosphere in winter.

LABITZKE, jointly with H. VAN LOON embarked in a systematic study extending the region under analysis toward lower latitudes, down into tropospheric altitudes, and to the other seasons. Concentrating on the height of the 30 mb level as an important indicator of the integral behaviour of the air column below, they identified a clear geographic dependence of the correlation with solar activity; in particular, during mid-winter and east years of the QBO, the basic correlation pattern exhibits a crescent-shape region of positive correlation of the 30 mb level height along 30°N over the Pacific, with a pattern of anti-correlation centred over the Arctic. During the west years, the only important correlation (positive) is found over the Arctic. A consistent average geographic pattern of correlation with solar activity subsists for all data regardless of the QBO phase, as shown in Fig. 5 (LABITZKE & VAN LOON, 1993).

Figure 6 shows the average change of temperature from solar maximum to solar minimum as a function of geopotential height for Lihue (Hawaii), which is situated under the maximum correlation area of Fig. 5. Different epochs during the year are shown (the reversal at the tropopause at about 100 mb is an inherent property of the atmosphere). The annual mean temperature difference in the troposphere between solar maximum and minimum is substantial, 1.8°C at 300 mb; even at the surface it is 0.9°C in summer. The large positive correlation between 30 mb height and solar flux in the Pacific region is thus mainly due to the difference in tropospheric temperature between the extremes of the 11-year sunspot cycles; this is significant for the identification of trigger mechanisms responsible for solar-climate relationships (see Chapter 5). As to the aliasing effect mentioned in "freak fact N° 4", it does not apply to a strongly period-modulated variation such as the QBO (TINSLEY & HEELIS, 1993).

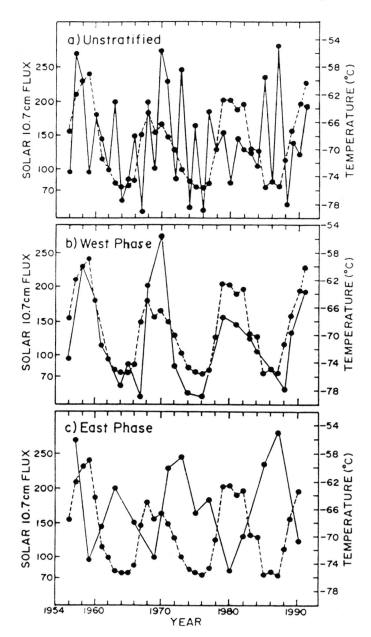

Fig. 4   10.7 cm solar radio wave flux (broken line--in units of 10^{-22} W/m^2 Hz) and the mean 30 mb temperature at the North Pole during winter (January-February). (a) All years; (b) Years with west phase of the QBO (westerly winds at the equatorial 50-40 mb level during January-February); (c) Years with east phase (easterly winds). Updated from LABITZKE (1987)

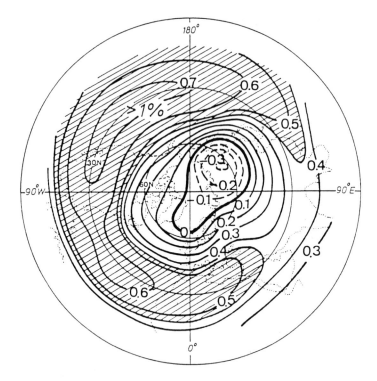

Fig. 5 Contours of equal correlation coefficient between the annual mean height of the 30 mb level (1958-1992) and the 10.7 cm solar flux (hatched area: local statistical significance greater than 1%). From LABITZKE & VAN LOON (1993)

An intriguing result is that of MENDOZA et al. (1991) concerning statistics of the occurrence of El Niño events from 1700 to 1985. It had been noted earlier (PÉREZ-ENRÍQUEZ et al., 1989) that the most intense events have occurred during periods of anomalous solar activity and that in general El Niño-Southern Oscillation (ENSO) events tend to gather around peaks of auroral activity during the descending phase of solar sunspot cycle. MENDOZA et al. (1991) analysed the frequency of occurrence of ENSO events according to intervals of sunspot gradient values (time derivative expressed in sunspot number change per year). They conclude that 63% of the ENSO events occur during the descending phase of the solar cycle, and that there are twice as many events occurring one year after the maximum rate of sunspot decline (maximum negative gradient) than what would be expected by random occurrence (see Fig. 7). While at least part of the first result can be explained by the asymmetry of the sunspot number curve with respect to each maximum (notice in Fig. 3 that the sunspot number rise is usually faster and of shorter duration than the decay), the second result is more significant. Indeed, the authors conclude from a computer experiment that the probability for the peak in Fig. 7 to occur by chance is less than 0.4 %.

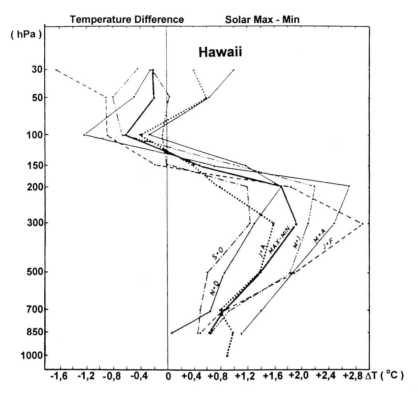

Fig. 6 The average change of temperature from solar minimum to maximum as a function of atmospheric depth, for Lihue (Hawaii), through the year and in the annual mean (heavy line), for a 36-year series. From LABITZKE & VAN LOON (1993)

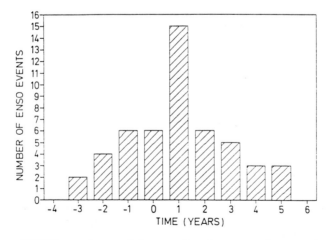

Fig. 7 Histogram of El Niño events around the year of maximum rate of sunspot decrease during the descending phase of the solar cycle, for the period 1700-1985. From MENDOZA et al. (1991)

4.2 Long-term variations

The annual average northern hemisphere land air temperature curve of JONES et al. (1986) (or HANSEN & LEBEDEFF (1987) for the global temperature), popularly known as the "global warming curve", is usually interpreted by non-scientists (and the media) as being entirely due to an anthropogenically enhanced greenhouse effect. However, it is clear that not all of the global temperature variation in the last 100 years could have been of anthropogenic origin: (1) most of the increase during this century (78%) took place before 1940 when the rate of increase of CO_2 was much lower than at present; (2) there was a steady temperature decrease between 1940 and the early seventies while the CO_2 kept increasing at an accelerated rate; (3) the global temperature had already been rising during the two previous centuries since the Little Ice Age. Still, it came as a surprise when FRIIS-CHRISTENSEN & LASSEN (1991) published a paper on the correlation between the global temperature and the length of the solar cycle, which at first sight seemed to indicate that the entire temperature behaviour during the last 100 years could be due to solar variability (although the authors never stated this - see "freak fact N° 6").

Recently, FRIIS-CHRISTENSEN & LASSEN (1993) extended this correlation back to 1750 using northern hemisphere temperature data by GROVEMAN & LANDSBERG (1979). They also re-computed the solar cycle length curves using a filtering procedure to take into account earlier criticism of the determination of solar cycle length as a time-dependent parameter. As Fig. 8 shows, the remarkable relationship between average annual temperature and solar cycle length persists. In their 1991 paper, the authors noted the similarity of the global temperature anomaly and the 11-year running mean of the sunspot number, but pointed out the fact that the temperature curve was leading the sunspot number curve by up to 20 years, which of course ruled out any causal connection between the two. But as shown in Fig. 8, this time shift disappears when sunspot cycle length is used, implying that it is the cycle length and not the sunspot number that appropriately represents the climate-relevant part of solar variations (for instance, the changes in irradiance). Both the 11-year running mean sunspot number and cycle length do track each other well, but the cycle length leads by 10-15 years (see "freak fact N° 3").

Another study of long-term correlations was conducted by REID (1991), who used the sea surface temperature record as a climate change indicator. Despite some questions about quality of data before the beginning of the century, global average sea surface temperature variations have certain advantages: because of the thermal inertia of the ocean, they do not exhibit short-term variations as the land temperatures; they do not have to be corrected for effects from recent urban growth; they exhibit greater spatial coherence; and they represent samples from an area covering 70% of the earth's surface. As mentioned in the previous Chapter, REID derived an empirical relationship between solar irradiance and sunspot number and, using a one-dimensional columnar model of the coupled atmosphere-ocean system, computed the theoretical average sea surface temperature, shown in Fig. 9 together with the experimental data. The calculation consisted of the integration of a heat diffusion equation in the ocean column governed by a set of coefficients and subjected to given coupling and boundary conditions at the top and bottom of the ocean, respectively; this equation was integrated forward in time from the late 1600s, and is shown to fit well the measured average temperature variations (Fig. 9).

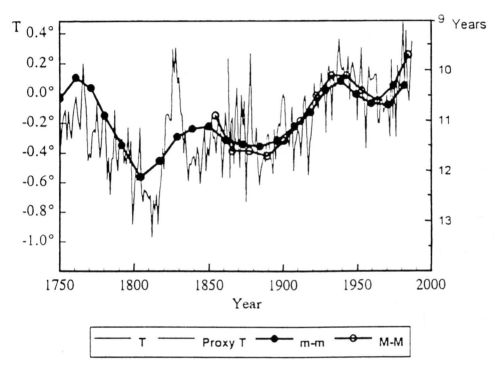

Fig. 8 Annual average values of the northern hemisphere temperature (thin curve) and appropriately filtered values of the sunspot cycle length (solid curves, determined independently by means of sunspot minima (m-m) and maxima (M-M); note reverse scale). See FRIIS-CHRISTENSEN & LASSEN (1993)

As REID pointed out, the experimental average sea surface temperature follows in general lines the global land temperature, but it is delayed with respect to the latter by 10-15 years. This is three times the value of the time constant for radiative equilibrium of the ocean mixed layer used by REID in the model calculation. It is reasonable to assume that if one were to use the solar cycle length, which precedes the envelope of the sunspot number by 10-15 years (see "freak fact N° 3"), one would have to use a larger (and perhaps more realistic) value of the ocean response time to achieve a good fit. Another point worth re-examining is the unrealistically large secular variation of solar irradiance used in REID's calculation (Chapter 3); an appropriate amplification mechanism might have to be introduced on the atmospheric side of the model (see next Chapter).

Finally, concerning very long-term changes, ANDERSON (1992) reported on a possible connection between surface winds, solar activity and the earth's internal magnetic field. The author has identified an association between an enhanced 100-200-year solar cycle periodicity (as revealed in the ^{14}C record in tree-rings) and fluctuations in surface wind intensity on a 200-year time scale (as revealed in the thickness of varved sediments in a Minnesota lake). This association existed only during the mid-Holocene 5000-7500 years ago, when the earth's magnetic dipole moment went through the last minimum, i.e., when the intensity of the solar-activity-modulated cosmic rays entering the atmosphere was at a maximum.

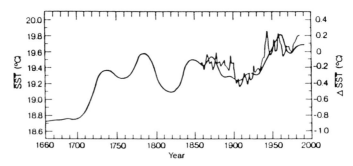

Fig. 9 Global average sea-surface temperature calculated from REID's model, and the observed time series (REID, 1991)

4.3 Short-term variations

The studies of possible effects of short-term solar-activity-related variations on the tropospheric system are on much less firm footing than the 11-year cycle and longer-term correlations, despite the fact that, in principle, the former would be easier to identify statistically than the latter because of the unique signatures of each one of the possible solar inputs on a day-to-day time scale. These studies are mostly isolated efforts by individual scientists, each one of which has chosen one given pair of solar-atmospheric variables out of the many possible ones; a concerted, internationally coordinated approach is still missing.

One example is SCHUURMANS' (1991) study of the temperature in the troposphere (500 mb level) and lower stratosphere (200 mb) over De Bilt, The Netherlands, and its behaviour during 72 solar proton events that occurred in the interval 1955-1984. The author finds that for the east phase of the QBO there is a clear reduction of the temperature at the 200 mb level of about -2.4°C that persists at least 3 weeks after the proton event. No measurable effect is seen during west QBO. Comparing the behaviour of the 200 and 500 mb levels, the author argues that the cooling in the lower stratosphere is not due to a dynamically related warming of the troposphere; rather, he postulates a solar-induced heat sink operating at 200 mb.

Another example is the work by PUDOVKIN & BABUSHKINA (1992) who studied atmospheric transparency variations during the interval 1961-1984 associated with geomagnetic disturbances, using monthly actinometric data from a network of meteorological observatories in three latitudinal bands of the former Soviet Union. They find a considerable increase in solar radiation intensity (increase in transparency) in the auroral zone, 1-2 days after the onset of a geomagnetic storm. The authors speculate that this may be due to chemical changes in the stratosphere as the result of storm-associated Forbush decreases of cosmic ray intensity.

5. Mechanisms

As stated in Chapter 2, because of energy balance considerations, the mechanisms responsible for solar variability effects on climate should be trigger-mechanisms which catalyse or control the release of latent energy, thus leading to a large-scale redistribution of energy in the atmospheric system. To explain the observed effects, one must look for amplification processes that respond by a factor of at least one million to small variations of energy inputs. There are three basic candidates for mechanisms, sketched in the lower part of Fig. 1 and discussed below.

5.1 Total irradiance variations

These variations correspond to the main photon channel at left in Fig. 1. Their 11-year variation is shown in Fig. 2; a bigger change can be expected on a long-term, secular scale (Chapter 3). LABITZKE & VAN LOON (1993) proposed the modulation of the Hadley cell circulation as a possible mechanism, based mainly on the fact that the middle and upper troposphere under the crescent-shaped region between 20°N and 45°N in the Pacific-Atlantic area shown in Fig. 5 is consistently warmer during solar maximum (e.g., see Fig. 6). This points to an enhanced Hadley circulation in which the troposphere in the Intertropical Convergence Zone is warmed more by an increased release of latent heat during solar maximum. The reversal of the temperature effect at and above the tropopause in Fig. 6 indicates that the primary enhanced heating process resides within the tropopause (because the vertical temperature difference profiles in Fig. 6 are indeed similar to the profiles of differences between non-solar related strong and weak Hadley circulation periods). The control by the QBO of wave and energy propagation characteristics to higher latitudes in the winter hemisphere (Section 4.1) would be responsible for the 11-year modulation of the polar stratosphere (Fig. 4). In the summer hemisphere, in absence of a well-defined polar vortex, the coupling between the equatorial and polar stratosphere is less complex, and the atmosphere exhibits a behaviour parallel to the 11-year cycle regardless of the phase of the QBO (VAN LOON & LABITZKE, 1990). This is yet another argument in favour of the Hadley circulation modulation hypothesis.

In principle, long-term solar variability effects on climate such as shown in Figs. 8 and 9 should be "easier" to explain because the input power variations could be expected to be several times larger than those found for the 11-year cycle (Chapter 3). However, it is not yet clear (although quite suggestive) whether the above Hadley cell modulation mechanism would also be applicable at this time scale. GCM model calculations do show that a total irradiance change of about 2 W/m^2 can have an effect on global temperature of about half that of doubling the CO_2 concentration.

5.2 Solar UV flux changes

Given that the UV flux powers the dynamics of the stratosphere *via* ozone absorption (e.g., HOOD et al., 1993), and given the large variability of the solar UV flux, it is reasonable to expect this flux to play a fundamental role in driving stratospheric variability (e.g., KODERA et al., 1991). Such role may be a contributing factor, but it could not explain other important findings by LABITZKE & VAN LOON, such as shown in Figs. 6 and 7. It may, however, be related to the behaviour of the polar stratosphere in winter during the west phase of the QBO, when its temperature is positively correlated with solar activity (Section 4.1).

Finally, we may speculate that the long-term variability of solar UV radiation may have an effect on climate *via* the response to UV of the phytoplankton in the oceans' euphotic zone, which is believed to play a fundamental role in the global control of ocean uptake and release of CO_2.

5.3 Atmospheric ionization and global electric circuit

The flux of galactic cosmic rays is the dominant source of continuous ionization in the troposphere and the lower stratosphere; because of the modulation of this flux by the interplanetary magnetic field, this ionization is variable. Fig. 1 depicts several possible solar-controlled ways in which the ionization can change. This ionization determines atmospheric conductivity, which in turn regulates the clear-air vertical electric field and electric current between the ionosphere and ground. The latter, at high latitudes, is magnetically connected to the solar wind and its electric potential is controlled by the IMF. The electrical connections in the system polar ionosphere/mid-latitude and equatorial ionosphere/atmosphere/ground is called the "global electric circuit". TINSLEY has proposed a mechanism (e.g., TINSLEY & HEELIS, 1993) connecting atmospheric electricity and the rate of contact ice nucleation in clouds, which can operate on a time scale of hours. This theory is based on the fact that the rate at which the charging of water droplets proceeds will depend on the vertical atmospheric current and therefore be responsive to both the cosmic ray flux and the local ionospheric potential. Tinsley's "electrofreezing" mechanism postulates that variations in the amount of such charge affect the rate of initial ice generation at the top of clouds, with ensuing effects on cloud formation.

6. Final thoughts

Achieving a scientific understanding of the inner workings of the terrestrial environment is one of the most difficult and ambitious endeavours of humankind, rivalling in complexity the harnessing of nuclear energy, the conquest of space, and the understanding of the human brain. Unfortunately, the more we learn about the environment, climate and anthropogenic effects, the more political problems emerge. Lawyers, judges and politicians are expected to render verdicts and make decisions on the basis of scientific concepts they grasp poorly and without a clear understanding of the scientific method, the inherent experimental uncertainties, and the natural limitations of scientific predictability.

One of the key tasks in climate research at the present time is to learn to separate anthropogenic effects from natural change. Solar variability is one possible cause of natural change in addition to other external phenomena such as volcanic eruptions and, on a long-term scale, orbital changes, celestial body impacts and tectonic plate motion. The unpredictable intrinsic variations of a highly non-linear system such as the atmosphere also must be counted as a "natural" change. Thus, before far-reaching policy decisions are made that in any way limit or regulate human activity, a scientific understanding of all possible external influences, however minor at first sight, is a condition *sine qua non* from a very practical point of view: without such knowledge there is the danger of imposing billion-dollar solutions for million dollar problems and face possible economic disaster, or finding million dollar solutions to billion-dollar problems and face possible environmental disaster.

Acknowledgments

The author was supported for his work by grant ATM 92-12638 from the Division of Atmospheric Sciences of the National Science Foundation and grant NAGW-1342 from the Space Physics Division of the National Aeronautics and Space Administration. An invitation from the European Science Foundation to present this review at the Workshop on Solar Output and Climate in the Holocene (Bologna, 1-3 April, 1993) was greatly appreciated.

References

ANDERSON, R. Y. (1992): Possible connection between surface winds, solar activity and the earth's magnetic field. Nature 358, 51-53

BAKER, D. N.; BLAKE, J. B.; GORNEY, D. J.; HIGBY, P. R.; KLEBESADEL, R. W. & KING, J. H. (1987): Highly relativistic electrons: a role in coupling to the middle atmosphere. Geophys. Res. Lett. 14, 1027

BUCHA, V. (1988): Influence of solar activity on atmospheric circulation types. Ann. Geophysicae 6, 513-524

CLAYTON, H. H. (1923): World Weather. Mac Millan, New York

DANSGAARD, W.; JOHNSEN, S. J.; CLAUSEN, H. B.; DAHL-JENSEN, D.; GUNDESTRUP, N. S.; HAMMER, C. U.; HVIDBERG, C. S.; STEFFENSEN, J. P.; SVEINBJÖRNSDOTTIR, A. E.; JOUZEL, J. & BOND, G. (1993): Evidence for general instability of past climate from a 250 kyr ice-core record. Nature 364, 218-220

DONNELLY, R. F. (1991): Solar UV spectral irradiance variations. J. Geomagn. Geoelectr. 43, Suppl., 835-842

EDDY, J. (1976): The Maunder Minimum. Science 192, 4245, 1189-1202

FRIIS-CHRISTENSEN, E. & LASSEN, K. (1991): Length of the solar cycle: an indicator of solar activity closely associated with climate. Science 254, 698-700

FRIIS-CHRISTENSEN, E. & LASSEN, K. (1993): Two and a half centuries of solar activity variations and a possible association with global temperature. Danish Meteorol. Inst. Scientific Report 93-4

GROVEMAN, B. S. & LANDSBERG, H. E. (1979): Simulated northern hemisphere temperature departures 1579-1880. Geophys. Res. Lett. 6, 767-769

HANSEN, J. & LEBEDEFF, S. (1987): Global trends of measured surface air temperature. J. Geophys. Res. 92, 13345-13372

HINES, C. O. & HALEVY, I. (1977): On the reality and nature of certain sun-weather correlation. J. Atmosph. Sci. 34, 382-404

HOOD, L. L.; JIRIKOWIC, J. L. & McCORMACK, J. P. (1993): Quasi-decadal variability of the stratosphere: Influence of long-term solar ultraviolet variations. J. Atmosph. Sci. 50, 3941-3958

JONES, P. D.; RAPER, S. C. B.; BRADLEY, R. S.; DIAZ, H. F.; KELLY, P. M. & WIGLEY, T. M. L. (1986): Northern hemisphere surface air temperature variations: 1851-1984. J. Clim. Appl. Met. 25, 161-179

KODERA, K.; CHIBA, M. & SHIBATA, K. (1991): A general circulation model study of the solar and QBO modulation of the stratospheric circulation during the northern hemisphere winter. Geoph. Res. Lett. 18, 1209-1212

KOSTERS, J. J. & MURCRAY, D. G. (1979): Change in the solar constant between 1968 and 1978. Geophys. Res. Lett. 6, 382-384

KYLE, H. L.; HOYT, D. V.; HICKEY, J. R.; MASCHOFF, R. H. & VALLETTE, B. J. (1993): NIMBUS-7 earth radiation budget calibration history. Part I: The solar channels. NASA Reference Publication 1316

LABITZKE, K. (1982): On the interannual variability of the middle stratosphere during northern winter. J. Meteor. Soc. Japan 60, 124-139

LABITZKE, K. (1987): Sunspots, the QBO, and the stratospheric temperature in the North Polar region. Geophys. Res. Lett. 14, 535-537

LABITZKE, K. & VAN LOON, H. (1993): Some recent studies of probable connections between solar and atmospheric variability. Ann. Geophysicae 11, 1084-1094

LEAN, J.; SKUMANICH, A. & WHITE, O. (1992): Estimating the sun's radiative output during the Maunder Minimum. Geoph. Res. Lett. 19, 1591-1594

MENDOZA, B.; PÉREZ-ENRÍQUEZ, R. & ALVAREZ-MADRIGAL, M. (1991): Analysis of solar activity conditions during periods of El Niño events. Ann. Geophysicae 9, 59-54

NESME-RIBES, E. (1995): The Maunder Minimum and the deepest phase of the Little Ice Age: a causal relationship or a coincidence? In: Frenzel, B.; Nanni, T.; Galli, M. & Gläser, B. (eds.): Solar output and climate during the Holocene. Paläoklimaforschung/ Palaeoclimate Research 16 (this volume), 131-144

PÉREZ-ENRÍQUEZ, R.; MENDOZA B. & ALVAREZ-MADRIGAL, M. (1989): Solar activity and El Niño: The auroral connection. Nuovo Cimento 12C, 223-230

PITTOCK, A. B. (1979): Solar cycles and the weather: successful experiments in autosuggestion? In: McCormac, B. M. & Seliga, T. A. (eds.): Solar-terrestrial influences on weather and climate, D. Reidel Publ. Co., Dordrecht, Holland, Boston, USA, London, England, 181-191

PUDOVKIN, M. I. & BABUSHKINA, S. V. (1992): Atmospheric transparency variations associated with geomagnetic disturbances. J. Atm. Terr. Electr. 54, 1135-1138

REID, G. C. (1991): Solar total irradiance variations and the global sea surface temperature record. J. Geophys. Res. 96, 2835-2844

RIND, D. & OVERPECK, J. (in press): Hypothesized causes of decadal-to-century climate variability: Climate model results. Quat. Sci. Rev. (in press)

ROBERTS, W. O. (1975): Relationship between solar activity and climate change. In: Bandeen, W. R. & Maran, S. P. (eds.): Goddard Space Flight Center Special Report NASA SP-366

ROEDERER, J. G. (1992): The Solar-Terrestrial Energy Programme 1990-1997. COSPAR Info. Bull. 123, 47-54

SALBY, M. L. & SHEA, D. J. (1991): Correlations between solar activity and the atmosphere: an unphysical explanation. J. Geophys. Res. 96, 22579-22595

SCHUURMANS, C. J. E. (1991): Changes of the coupled troposphere and lower stratosphere after solar activity events. J. Geomagn. Geoelectr. 43, 767-773

TEITELBAUM, H. & BAUER, P. (1990): Stratospheric temperature eleven-year variations: solar cycle influence or stroboscopic effect? Ann. Geophysicae 8, 239-242

TINSLEY, B. A. & HEELIS, R. A. (1993): Correlations of atmospheric dynamics with solar activity: Evidence for a connection *via* the solar wind, atmospheric electricity, and cloud microphysics. J. Geophys. Res. 98, 10, 375-10, 384

VAN LOON, H. & LABITZKE, K. (1990): Association between the 11-year solar cycle, the QBO and the atmosphere. Part IV: the stratosphere, not grouped by the phase of the QBO. J. Climate 3, 827-837

WHITE, O. R.; SKUMANICH, A.; LEAN, J.; LIVINGSTON, W. C. & KEIL, S. L. (1992): The sun in a noncycling state. Publ. Astron. Soc. Pacific 104, 1139-1143

WILCOX, J. M.; SCHERRER, P. H.; SVALGAARD, L.; ROBERTS, W. O.; OLSON, R. H. & JENNE, R. L. (1974): Influence of solar magnetic sector structure on terrestrial atmospheric vorticity. J. Atmosph. Sci. 31, 581-588

WILLSON, R. C. & HUDSON, H. S. (1988): Solar luminosity variations in solar cycle 21. Nature 332, 810-812

Author's address:

Prof. J. G. Roederer, Geophysical Institute, University of Alaska Fairbanks,
Fairbanks AK 99775-7320, USA

Solar activity signal in the mean winter temperatures

Stefano Cecchini, Menotti Galli & Teresa Nanni

Summary

The mean seasonal and annual temperatures of continental Europe and western Siberia are analysed in order to detect possible solar signals. It is found that winter temperature variability largely dominates the annual mean. We suggest that this fact might depend on variations in the solar constant. Furthermore the winter temperatures show significant quasi biennial variations (2.4 years) correlated with similar variations of the solar activity index R_z. The cyclic variation of winter temperatures on a time scale of 12 years appear significantly related to the 10.7-year mean periodicity of the index R_z during the 1880-1980 interval. These last two possible signals appear to be independent of solar irradiance; they may instead be related to ultraviolet emission and to its influence on the general circulation by some triggering effect.

Résumé

On a cherché la présence d'un signal solaire dans la température moyenne saisonnière et annuelle de l'Europe Continental et de la Sibérie Oriental. La moyenne de la température annuelle est determinée par la variabilité de la temperature de l'hiver, qui peut être determinée par la variation de la constante solaire. De plus la température de l'hiver montre aussi des oscillations presque biennales (2.4a), qui présentent une correlation avec la variation de l'index R_z de l'activité solaire. Les variations de la température hivernal en cycles de 12 années montrent une corrélation significative avec la périodicité moyenne de 10.7 années de l'index R_z pendant la période 1880-1980. Ces deux signaux paraissent n'avoir pas de dépendance de l'irradiance solaire. Ils pourraient être reliés à l'émission ultraviolet et à sa influence sur la circulation générale par quelque "triggering" effet.

1. Introduction

The presence of a clear and unmistakable effect of solar activity on climatic phenomena is still a matter to be established. However the best established relationship observed so far is that reported by LABITZKE & VAN LOON (1990) concerning the correlation between the winter stratospheric temperatures of the northern hemisphere and sunspot cycles during 1956-1988. This correlation was achieved by selecting the years according to the phase of

east or west zonal stratospheric winds at the equator, connected with quasi biennial oscillations (QBO). The same authors (VAN LOON & LABITZKE, 1990) found high correlations for
not selected years between 10.7 cm solar flux and stratospheric heights of 100 and 30 hPa
in the northern hemisphere corresponding to medium latitude regions including Atlantic
and Pacific regions, during 20-25 years. Here we briefly report some results that may be
regarded as further evidence for a solar signal in atmospheric variables during the period
1880-1980.

2. The temperature data

Some authors have suggested that the Maunder Minimum of solar activity which extended
over the period 1645-1715 (EDDY, 1977), with the lowest solar activity values during 1690-
1700, may be related to the coldest period of the so-called Little Ice Age, which took place
at around the same period (LEGRAND et al., 1993). However when considering temperature
estimates either from measurements of tree-ring width during that period or from "vintage
dates" (PFISTER, 1992; LEGRAND, 1979) which are related to the vegetation season of
woody plants, this time interval cannot be considered particularly cold. On the other hand,
if one looks at the estimates of the cold season temperatures as obtained from historical
documents (LEGRAND et al., 1993; PFISTER, 1992), one can see that during that same
period, winter temperatures have actually reached one of the lowest records (Fig. 1) since
the sixteenth century.

For this reason and after considering observations on the effects of winter temperature on
the growth of the Italian stone pine (*Pinus pinea* L.) from Ravenna pine forest (GALLI et al.,
1992) we established the notion that changes in solar flux would probably be more effective in changing winter temperatures. We therefore decided to compare growth ring series
of Ravenna pine between 1880 and 1980 to mean winter temperatures for Bologna and to
the analogous mean winter temperatures of boxes 9 and 10 computed by HANSEN &
LEBEDEFF (1987). These temperatures correspond to the high latitude winter temperatures
(HLWT) established for continental Europe and Siberia above 45° latitude.

The first observation was that the mean winter temperature variations were much higher
than those of mean summer temperatures, so that the variation in mean annual temperature
strongly reflects the behaviour of winter temperature (Fig. 2). If one considers r.m.s. (root
mean square) deviations of annual and seasonal means, one finds that deviations of winter
means account for about 90% of the deviations of the annual means. Furthermore the
HLWT for the years 1880-1980 shows a strong quasi biennial variation. It means that for
most cases two subsequent minima or maxima are separated by one or two intermediate
values, whereas the mean high latitude summer temperature (HLST) shows longer term
variations. This is clearly shown by the power spectrum of HLWT (Figs. 3, 4) obtained
using the method of BLACKMAN & TUKEY (1958) with 32 lags and D3 lag window. For

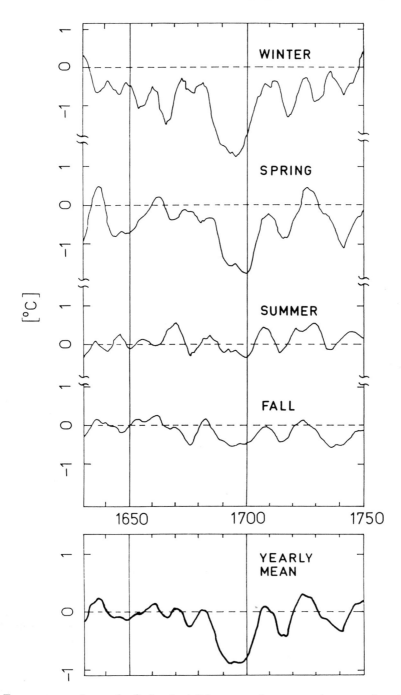

Fig. 1 Temperature estimates for Switzerland (11-year moving averages) expressed as departures from the mean for 1901-1960, adapted from PFISTER, 1992

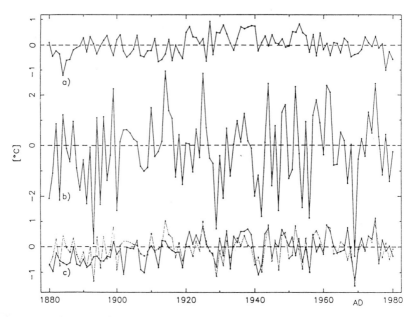

Fig. 2 Summer (a), winter (b) and annual (c) mean temperatures - all in the same temperature scales. Notice that winter variance is about 12 times that of the summer and that winter variation accounts for 90% of the annual variation. Winter and summer variations are independent at a high level of significance. Most variations in HLWT are biennial whereas in HLST are of longer periods

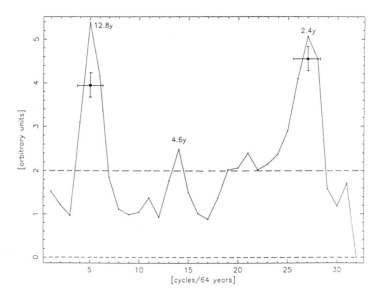

Fig. 3 The power spectrum of HLWT computed through the autocorrelation function multiplied by D3 Blackman and Tuckey window. Standard errors of the two most important bands are shown

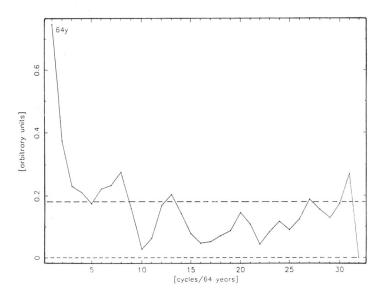

Fig. 4   Same as Fig. 3 but for HLST. The significant large peak at 1c/64 years is due to the drift of the series

HLWT very significant peaks are centred at the frequencies 1c/2.4 years = 1c/28 months, 1c/12.8 years, whereas the analogous spectrum of HLST shows no special dominant periodicity. The high value of the spectrum at 1c/64 years is interpreted as reflecting the trend of the series.

3. The solar signal in the temperature data series

In order to discover a possible link between winter temperature and the solar activity index Rz, one option is to look for a significant dependence between selected values of R_z and the corresponding temperature values. This can be done with the so-called method of "epoch superposition". In one of the two series (e.g., in the R_z index series) all the years (called "zero" years or "reference years") corresponding to a certain anomaly are selected. The mean value of the data in the other series corresponding to the same point in time are then computed and compared to the reference value. If a significant correlation between the two series exists for the chosen years, the computed mean will be significantly different from the mean obtained choosing the reference years with a random criterium.

The frequency bands we considered first were the ones in the range of one cycle per 2-3 years that appear in the winter temperatures. In order to amplify such variations we have computed the residuals R'_z of the R_z annual means from a moving binomial filter with coef-

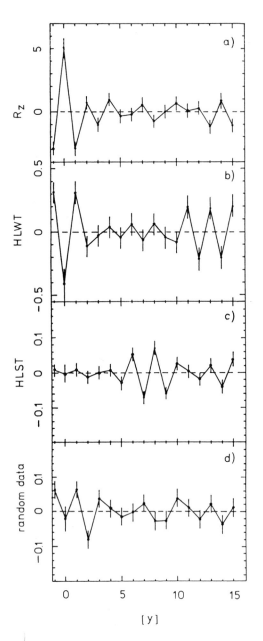

Fig. 5 Average superposed epoches of HLWT each corresponding to a relative maximum R_z for (a) R_z; (b) HLWT; (c) HLST; (d) random series. (see text). (Horizontal dashed lines mark the r.m.s. error of the folded epoches around their averages)

ficients (0.25, -0.5, 0.25). Then we chose the relative maxima of R'_z as reference years and averaged correspondingly not only the values of HLWT of the same years, but also the 15 preceding years and the 15 following years. We also carried out the same averaging operation on the R'_z data series using the same reference years. Corresponding to the expected high mean values of R'_z with respect to the remaining 30 values, we obtained a mean value of HLWT significantly lower than the 30 surrounding remaining values. As the values of the mean epoches of HLWT and R'_z to the right and to the left at the "0" years do not appear significantly different, we have folded and averaged the mean epoches around their central point.

When doing the analogous calculations for the minima of R'_z, we obtained similar results with a significant maximum of the mean epoch of HLWT at the "0" point and no significant differences between the left and right side of the same epoch. Thus averaging with their weights the folded epoches of R'_z maxima and HLWT, with the inverted and folded epoches of R'_z minima and HLWT, we obtained the curves a) and b) of Fig. 5. The same results of the method applied to the HLST series are presented in the graph c) of Fig. 5. The results of a simulation performed by using a series of random data with the same number of data, uniformly distributed and with the same variance as HLWT, are presented in graph d) of Fig. 5. The r.m.s. deviations for each point computed in Fig. 5 are shown. They clearly demonstrate the significance of the negative variation of HLWT with respect to the peaks of R'_z and on the independence between HLST and R'_z.

The enhanced oscillations after about 12 years from the "0" year (see Fig. 5b) might indicate a recurrence of higher or lower values of HLWT after an interval of 11-15 years. Alternatively, they can be caused by a wave consisting of a period of about 2 (1+1/13) years 26 months, coinciding with the peak of 26 months of the HLWT p.s.d. (power spectral density) of Fig. 4 and also found in the solar neutrino variation (ATTOLINI et al., 1988) and geomagnetic field intensity (SUGIURA & POROS, 1977).

In order to search for a further possible solar cycle signature on HLWT, we again used the method of superimposed epochs, taking the three values around each minimum as reference years and alternatively around each maximum of R_z and centring them in a time span of 31 years (Fig. 6a,b).

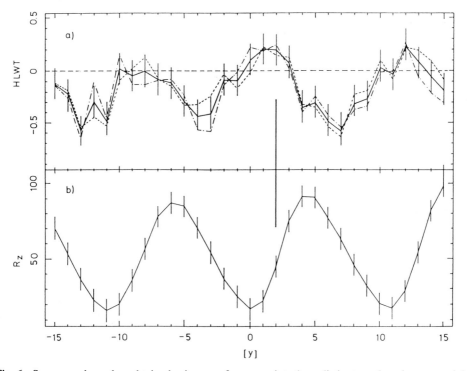

Fig. 6 Superposed epochs obtained using as reference points three distinct yearly values around the maxima or minima of each cycle of R_z for: (a) R_z reference minima averaged with reference maxima epoch after shifting it in such a way as to have minima and maxima coinciding; (b) mean HLWT epoch using: R_z minima (dotted line), R_z maxima shifted (dashed line), average of the two epochs (continuous line). Standard errors of superposed epochs are shown

When considering the minima of R_z as reference points, we obtained a mean epoch of HLWT shown in Fig. 6b with the dotted line, which has its maximum half way of the rising phase of the mean solar cycle. When, however, we take the maxima of R_z, compute mean epoches for R_z and HLWT, and shift them both to the right as to coincide with the R_z mean maxima epoches, we obtain the dashed curve in Fig. 6a. When the averages of the two HLWT superposed epoches (as shown in Fig. 6a) were then computed, the curve drawn as a continuous line was obtained, which shows a marked and significant correlation between R_z cycles and HLWT variation. In other words, a minimum of HLWT is to be expected half way up the ascending phase of a solar cycle.

The R_z-HLWT cross correlation function with its Fisher's significance values shown in Fig. 7 confirms the cyclic correlation between the two variables; its cross-spectrum of Fig. 8, computed with the lag window D3 and 32 lags, shows a significant peak corresponding to a mean cycle of 10.7 years also typical for R_z for the same time interval. As the cross spectrum has been obtained from the Fourier transformation of the estimates of the correlation coefficient function and therefore normalized to their standard deviation estimate, the spectrum of Fig. 8 is not affected by the relative values of their variations and, as it would appear in the coherence spectrum, between the two variables (JENKINS & WATTS, 1968).

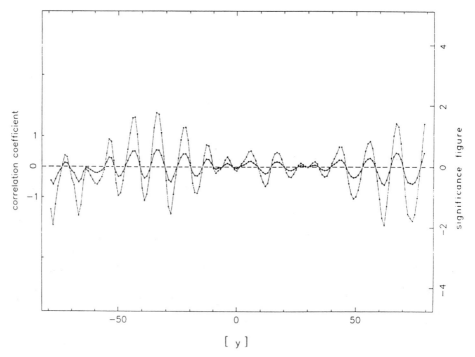

Fig. 7   The R_z-HLWT cross correlation function (continuous line) with its Fisher's significance value

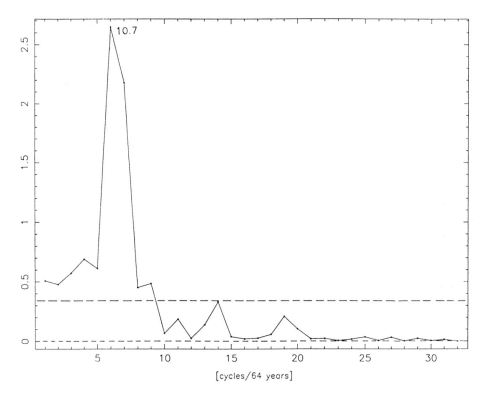

Fig. 8   The cross spectrum of R_z and HLWT computed by the Fourier transformation of the cross correlation function multiplied by D3 Blackman and Tukey window

4. Discussion

When considering the winter temperatures of the mentioned boxes 9 and 10, corresponding to those of continental Europe and western Siberia, we find some clues about the influence of possible signatures of the solar emission variation. Firstly, we consider the variation in winter temperatures which strongly affect the variations of annual mean. During the Maunder Minimum very low temperatures occurred particularly during winter time. These low temperatures were confirmed by glacier advances (LE ROY LADURIE, 1967) although these show no temperature anomalies in the summer time. The advances and retreats of glaciers, which reflect the annual mean temperature changes (WIGLEY, 1987), have been found to be directly correlated to carbon isotope production which in turn is driven by solar activity (WIGLEY & KELLY, 1990).

Variations in solar irradiance estimated on the basis of satellite observations do not explain the temperature variations in the last two centuries (WIGLEY & KELLY, 1990; KELLY &

WIGLEY, 1990; Fichefet, pers. comm.). However if one assumes that a change in irradiance was caused by variations in the solar magnetic network as suggested by Foukal (FRÖHLICH, 1993) one could explain the low winter temperatures observed during the Maunder Minimum and the ones towards the initial and the final decades of the nineteenth century (LAMB, 1972).

In fact Fichefet (pers. comm.) has found in a model experiment that small variations in solar irradiance mainly affect winter temperatures at higher latitudes and have only minor effects on summer temperatures. With a decrease of the solar constant by 0.25 % (about 3.5 W/m^2), as observed during the Maunder Minimum, he noted a decrease in winter temperatures at latitudes higher than 50°-60° and no significant summer temperature change as shown in boxes 9 and 10. If this can be generalized, we may conclude that variations in solar irradiance mostly affect winter temperatures. In time intervals of 2-3 years short rises in R_z seem to be associated with lower than average winter temperatures, and *vice versa* for negative variations of R_z. The effects appear significant when we carried out a similar computation and substituted random data for our summer or winter temperatures.

A more convincing relationship seems to exist between sunspot cycles and winter temperatures. This becomes evident when epoch superposition techniques are applied and it is found to be very significant when we compute the correlation function between the solar activity index R_z and smoothed HLWT with subsequent computation of its Fourier analysis. These two latter correlations favour the hypothesis that triggering effects, on some atmospheric circulation feature, can take place during the winter. These effects are thought to relate to rapid sunspot variations, which mostly happen during the development of big sunspot groups. These occur more frequently during the ascending phase of a solar cycle (KUIPER, 1953).

5. Conclusions

The analysis of mean seasonal temperatures for the boxes 9 and 10 by HANSEN & LEBEDEFF (1987) during 1880-1980 presents some aspects that appear to be related to the variability of solar output.

1) First of all it is interesting to note that winter variability is much higher than summer variability and that winter variation contributes about 90% to the variability of the annual mean.
2) Temperature variations present significant bands peaking around 2-3 years and 12 years. Peaks and troughs of temperatures are inversely related to peaks and troughs in short-term variations of the solar activity index R_z. Maxima and minima of sunspot cycles are significantly related: the temperature maxima are located at the ascending

phase of the R_z index and the mean solar cycle variations significantly affect the temperature epoch series.

3) According to a model experiment by Fichefet (pers. comm.) a variation in solar irradiance rate can affect temperature variations at latitudes higher than 50-60° during wintertime and no significant variation at any latitude during summertime. This agrees with what we have observed.

4) The phase and the possible temperature effect of solar activity for 12-year and 2.4-year time scales might also be related to the ultraviolet variation emitted by very active spot groups and to some kind of triggering effect on the general circulation pattern of the atmosphere.

References

ATTOLINI, M. R.; CECCHINI, S.; CINI CASTAGNOLI, G. & GALLI, M. (1988): On a biennial variation of the solar neutrino flux. Astro. Lett. and Communications 27, 55-61

BLACKMAN, R. B. & TUKEY, J. W. (1958): The measurement of power spectra. Dover Publ. Inc., New York, 190 p.

EDDY, J. A. (1977): The solar output and its variations. Ed. by O. R. White, Colorado Ass. Univ. Press, 51 p.

FICHEFET, T. (1993): pers. comm.

FRÖHLICH, C. (in press): Changes in total solar irradiance. J. Geophys. Res. (in press)

GALLI, M.; GUADALUPI, M.; NANNI, T.; RUGGIERO, L. & ZUANNI, F. (1992): Ravenna pine trees as monitors of winter severity in northeast Italy. Theor. Appl. Climatol. 45, 217-224

HANSEN, J. & LEBEDEFF, S. (1987): Global trends of measured surface air temperature. J. Geophys. Res. 92, D11, 13345-13372

JENKINS, G. M. & WATTS, D. G. (1968): Spectral analysis and its applications. Ed. by Jenkins, E. & Parzen, E., Holden-Day, San Francisco, 525 p.

KELLY, P. M. & WIGLEY, T. M. L. (1990): The influence of solar forcing trends on global mean temperature since 1861. Nature 347, 460-462

KUIPER, G. P. (1953): The Sun. Ed. by G. P. Kuiper, Chicago, Illinois

LABITZKE, K. & VAN LOON, H. (1990): Association between the 11-year solar cycle, the quasi biennial oscillation and the atmosphere: a summary of recent work. Phil. Trans. Roy. Soc. London A 330, 577-589

LAMB, H. H. (1972): Climate: Present, Past and Future. Methuen, London, 113 p.

LEGRAND, G. P. (1979): L'expression de la vigne au travers du climat depuis le moyen âge. Revue française d'oenologie 75, 23-50

LEGRAND, J. P.; LE GOFF, M. & MAZAUDIER, C. (1990): On the climatic changes and the sunspot activity during the XVIIth century. Annales Geophysicae 8/10, 634-644

LE ROY-LADURIE, E. (1967): Histoire du climat depuis l'an mil. Paris, Editions Flammarion, 366 p.

Pfister, C. (1992): Monthly temperature and precipitation in central Europe 1525-1979: quantifying documentary evidence on weather and its effects. In: Bradley, R. S. & Jones, P. D. (eds.): Climate Since A.D. 1500. Routledge, London

Sugiura, M. & Poros, D. J. (1977): Solar-generated quasi-biennial geomagnetic variation. J. Geophys. Res. 82, 5621

van Loon, H. & Labitzke, H. (1990): Association between 11-year solar cycle and the atmosphere. Part IV: The stratosphere, not grouped by the phase of the QBO. J. Climate 3, 827-837

Wigley, T. M. L. & Kelly, P. M. (1990): Holocene climatic change, ^{14}C wiggles and variations in solar irradiance. Phil. Trans. Roy. Soc. London A 330, 547-560

Wigley, T. M. L. (1987): The climate of the past 10,000 years and the role of the sun. In: Stephenson, F. R. & Wolfendale, A. W. (eds.): Secular solar and geomagnetic variations in the last 10,000 years. Kluwer, 209-224

Authors' addresses:

Dr. S. Cecchini, Istituto TESRE-CNR, Via Gobetti 101, I-40126 Bologna
Prof. M. Galli, Dipartimento di Fisica, Università di Bologna, Via Irnerio 46, 40126 Bologna
Dr. T. Nanni, Istituto FISBAT-CNR, Via Gobetti 101, I-40126 Bologna

Greenland palaeo-temperatures derived from the GRIP ice core

Willi Dansgaard, Sigfus J. Johnsen, Henrik B. Clausen, Niels Gundestrup, Claus U. Hammer & Henrik Tauber

Summary

A new 3029 m long deep ice core to bedrock has been drilled at the very top of the Green-land ice sheet (Summit) by the Greenland Ice-Core Project (GRIP), a European joint effort organized by the European Science Foundation. The ice core reaches back to 250 kyr B.P. according to dating based partly on stratigraphic methods and partly on ice-flow modelling. A continuous and extremely detailed stable isotope ($\delta^{18}O$) profile along the entire core depicts dramatic temperature changes in Greenland through the last two glacial periods, according to a time scale calculated by ice-flow modelling. Prior to 11 kyr B.P. the record is dominated by abrupt and irregular shifts, corresponding to temperature changes of 5° to 10°C, under glacial and maybe also under interglacial conditions. No less than 24 glacial interstadials have been identified in Weichselian ice, some of them are tentatively corre-lated to European interstadials. The shifts are associated with changes of the atmo-sphere/ocean circulation pattern in the North Atlantic Ocean. The Eem/Sangamon inter-glacial apparently elapsed completely differently in Greenland and Europe, perhaps because the Greenland climate was influenced by a strong and varying East Greenland Current, whereas Europe was permanently shielded by a strong Gulf Current displaced closer to western Europe than now.

Zusammenfassung

Ein neuer 3029 m langer Eiskern wurde vom Greenland Icecore Project, einer von der European Science Foundation koordinierten, europaweiten Initiative, vom höchsten Punkt des grönländischen Inlandeises (Summit) bis zum Felsbett gebohrt. Datierungen, die sich teils auf stratigraphische Methoden und teils auf Eisfließmodellierung stützen, weisen darauf hin, daß der Eiskern bis 250.000 J.v.h. zurückreicht. Eine fortlaufende und sehr detaillierte Meßreihe des stabilen Sauerstoffisotops ($\delta^{18}O$) entlang des gesamten Kernes zeigt, laut einer auf Eisfließmodellierung gegründeten Zeitskala, dramatische Tempe-raturschwankungen während der letzten beiden glazialen Perioden. In der Zeit vor 11,000 J.v.h. ist die Meßreihe durch abrupte und irreguläre Sprünge gekennzeichnet. Diese Sprünge entsprechen einem Temperaturunterschied von 5° bis 10°C unter glazialen und vielleicht auch unter interglazialen Bedingungen. Nicht weniger als 24 glaziale Zwischen-stufen wurden im Eis der Weichselkaltzeit identifiziert und einige davon werden versuchsweise mit europäischen Interstadialen korreliert. Die Temperatursprünge hängen mit Änderungen im atmosphärischen/ozeanographischen Zirkulationsmuster im Nordatlantik zusammen. Die Eem/Sangamon Warmzeit verlief anscheinend ganz unterschiedlich in Grönland und in Europa, was möglicherweise daran liegt, daß das Klima in Grönland von einer starken und wechselnden ostgrönländischen Strömung beeinflußt wurde, während sich Europa ständig unter dem Einfluß eines starken Golfstroms, der im Verhältnis zu heute näher zur Küste Westeuropas hin verschoben war, befand.

1. Introduction

The main objective of ice core drilling and analysis is to study climate and climate changes. Ice cores from polar ice sheets offer great potential for such studies, because each annual layer contains samples of the snow and the atmosphere with their impurities at the time when the layer was formed: the isotopic composition of the ice reveals information about the temperature (DANSGAARD et al., 1973; JOHNSEN et al., 1989); dust about the storminess (HAMMER et al., 1985); air bubbles about the greenhouse gases in the atmosphere (OESCHGER et al., 1984); acidity about volcanic eruptions in the northern hemisphere (HAMMER, 1980), etc. Taken together, they provide a detailed picture of the environment and climate of the past. All these parameters can be followed year by year and thus throw light on various facets of the complicated mechanism of climate.

In 1966 the USA Cold Regions Research Engineering Laboratory drilled the first deep ice core at Camp Century in northwest Greenland (Fig. 1), where the ice turned out to be 1390 m thick (UEDA & GARFIELD, 1968) and 130,000 years old near the bottom (DANSGAARD et al., 1982). The second deep-drilling project began in Greenland, now as an American-Danish-Swiss collaboration (GISP 1; LANGWAY et al., 1985), at the Dye 3 radar station in south Greenland (1979-81, Fig. 1), using a Danish ice core drill, ISTUK, built on new principles (GUNDESTRUP & JOHNSEN, 1985). It reached bedrock 2037 m below the surface and was then ready for deep-drilling at Summit (Fig. 1), the highest point of the ice sheet, and the most promising drill site in the northern hemisphere.

2. The organization of the Greenland Ice-Core Project (GRIP)

Under the auspices of the European Science Foundation (ESF) a collaborative effort was initiated in 1989 with contributions from Belgium, Denmark, France, Germany, Iceland, Italy, Switzerland, the United Kingdom, and the Commission of European Communities. The aim was to drill and to analyse an ice core to bedrock at Summit, 72°34'N, 37°37'W, 3230 m above sea level, where the annual mean air temperature is -32°C. A steering committee with representatives of the ESF and the main contributors coordinated the scientific work of the GRIP participants to ensure full coverage without duplication. Thus, a research programme was set up based on international collaboration across the board.

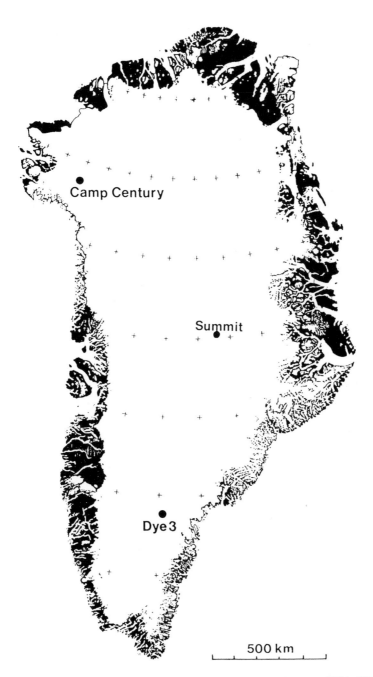

Fig. 1 Three deep drilling sites on the Greenland ice sheet: Camp Century (USA CRREL, 1966); Dye 3 (GISP1, 1981); and Summit (GRIP, 1992). In 1993 a fourth drilling to bedrock was completed 30 km west of Summit (GISP2)

The steering committee established the GRIP Operations Centre at the Geophysical Institute, University of Copenhagen, with a Greenland headquarters in Søndre Strømfjord airport, 800 km southwest of Summit.

3. Ice movements

As the annual snow layers are compressed into solid ice they sink under the weight of the succeeding layers. The layers are gradually stretched and get thinner and thinner as they approach the bedrock (Fig. 3). The stretching inevitably means horizontal movement, and the deepest ice follows all irregularities of the bedrock. It is this irregular movement that may disturb the stratification, as happened at Dye 3.

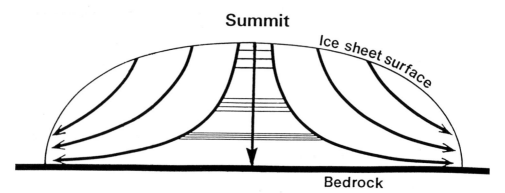

Fig. 2   Vertical cross section of an idealized ice sheet resting on a flat bedrock. The ice movement is symbolized by heavy arrows, the annual layers by horizontal lines. As they sink, the annual layers are stretched and therefore they get thinner and thinner - today their thickness is 23 cm of ice equivalent close to the surface at Summit, compared to a fraction of one mm at great depths. Only at Summit does the ice sink vertically

Only at one point on the ice sheet surface - at Summit - is there no horizontal movement. Only there does the ice sink vertically, so Summit offers the best chance of getting very far back in time in a continuous sequence, by drilling through the stack of layers.

4. The GRIP drilling

ISTUK (Gundestrup & Johnsen, 1985) is an 11 m long electromechanical drill. During the drilling it is suspended on a thin steel cable. The drillhole has to be filled with a thin oil of the same density as the ice to counteract pressure from the sides, which would otherwise quickly close the hole. Each time the drill comes up with 2.5 m of core (10 cm diameter), it is tilted to horizontal position for easy maintenance and core removal.

In the 1990 field season ISTUK reached a depth of 710 metres where the ice is 3840 years old. In 1991 it reached 2320 m where the ice is approximately 40,000 years old and on the 12th July 1992 drilling was stopped at a depth of 3028.8 metres because the cutting knifes were destroyed by hitting gravel and pebbles close to the bedrock. The six deepest metres of core were brownish with bedrock material.

5. Results

5.1 The temperature profile

A week after the drilling had terminated the temperature was measured along the hole. At the bottom it turned out to be -9°C, i.e. well below the melting point, and even considerably lower than the bottom temperature would have been, if the surface climate had always been the same as today. This is due to the coldness during the glaciation (DAHL-JENSEN & JOHNSEN, 1986). The postglacial heat wave that set in more than 11,000 years ago has not yet reached the deepest part of the ice sheet.

5.2 The stable isotope profile

The isotopic composition of polar glacier ice, $\delta^{18}O$ (hereafter denoted by δ[1]), is mainly determined by its temperature of formation, T °C (JOHNSEN et al., 1989):

$$\delta = 0.67 \, T - 13.7\text{‰.}$$

A continuous δ record along the upper 3000 m of the GRIP core is plotted on a linear depth scale in two sections of Fig. 3 (from DANSGAARD et al., 1993). The upper half of the record (A, δ scale on top) spans approximately the last 10 kyr, according to the time scale developed below. Each data point represents the mean δ value of a 2.2 m core increment, corresponding to the snow deposition of a few years near the surface, increasing to 20 years at 10 kyr B.P. Apart from the δ minimum at 1334 m depth, the record depicts a remarkably stable climate during the last 10 kyr. The postglacial climatic optimum would appear more noticeable if a δ correction turned out to be appropriate due to a possible decrease in surface elevation during the Holocene.

In contrast, the rest of the record shown in Fig. 3B is dominated by large and abrupt δ shifts, which suggests that climatic stability in Greenland is rare. Due to plastic thinning of the layers as they approach the bedrock, the lower 1500 m of ice represent a much longer

---

[1] $\delta^{18}O$ is the per mil deviation of the $^{18}O/^{16}O$ ratio in a sample from the $^{18}O/^{16}O$ value in SMOW (Standard Mean Ocean Water).

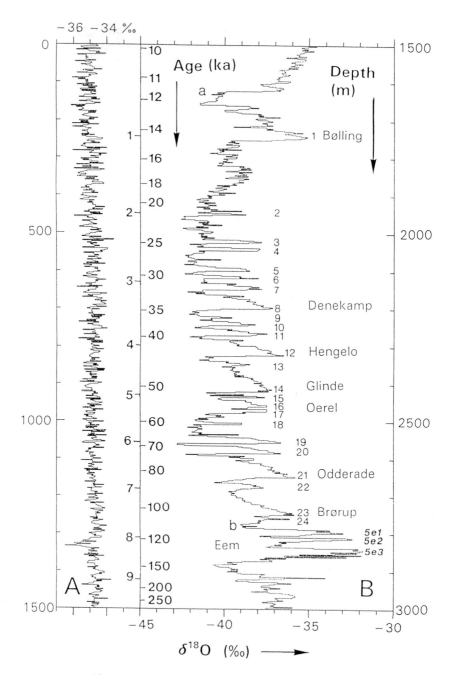

Fig. 3   The Summit δ^{18}O profile plotted on a linear depth scale (to the outer left and right). The two sections are separated by a time scale, which is established by counting annual layers back to 14.5 kyr B.P., and beyond that by ice-flow modelling (from Dansgaard et al., 1993)

period of time than the 1500 m above. On the right side of the record is an extension of the numbering of glacial interstadials (IS) introduced by JOHNSEN et al. (1992). Furthermore, a series of European pollen horizons (BEHRE & VAN DER PLICHT, 1992), ^{14}C dated back to 60 kyr B.P., is tentatively correlated with the longest lasting δ based interstadials.

5.3 The time scale

The two sections of Fig. 3 are separated by a time scale derived by counting annual layers downward from the surface back to 14.5 kyr B.P. (JOHNSEN et al., 1992), and further back in time by calculations based on ice-flow modelling:

$$t = \int_0^z dz / \lambda_z,$$

t being the age of the ice and λ_z the annual layer thickness at depth z. A steady state ice flow model (DANSGAARD & JOHNSEN, 1969) was modified by introducing a sliding layer at bedrock (JOHNSEN & DANSGAARD, 1992), and using a δ dependent accumulation rate (λ_H) derived from all available λ_z data:

$\lambda_H = \lambda_{H0} \exp [0.144 (\delta + 34.8)]$ m of ice equivalent per year,

λ_{H0} being the present value, 0.23 m/yr (JOHNSEN et al., 1992).

The other flow model parameters are as follows: the total thickness of the ice sheet H = 3003.8 m of ice equivalent; the thickness of the intermediate shear layer h = 1200 m; the ratio between the strain rates at the top of the silty ice and at the surface, $f_b = 0.15$; and the thickness of the silty ice layer, dh = 6 m. The latter value may be too low, but higher values would only influence the calculated ages of the deepest 50 m significantly, i.e. ice older than 250 kyr, which is not discussed in this paper.

The h and f_b values are chosen so as to assign well-established ages to two characteristic features in the δ record: (a) 11.5 kyr for the end of the Younger Dryas event (JOHNSEN et al., 1992; BARD et al., 1993), and (b) 110 kyr for the marine isotope stage (MIS) 5d (MARTINSON et al., 1987), which appear at depths of 1624 m and 2788 m respectively in the δ record, cp. points a and b in Fig. 3B. Back to 35 kyr B.P. the calculated time scale agrees essentially with that presented by DANSGAARD et al. (1993).

One of the assumptions behind the time scale calculation is that the stratigraphy has remained undisturbed, i.e. that all annual layers are represented in a continuous sequence and thinned according to the depth dependent vertical strain described by the flow model. This may fail at great depths due to folding close to a hilly bedrock and/or random thinning of layers of different rigidity (boudinage effect; STAFFELBACH et al., 1988).

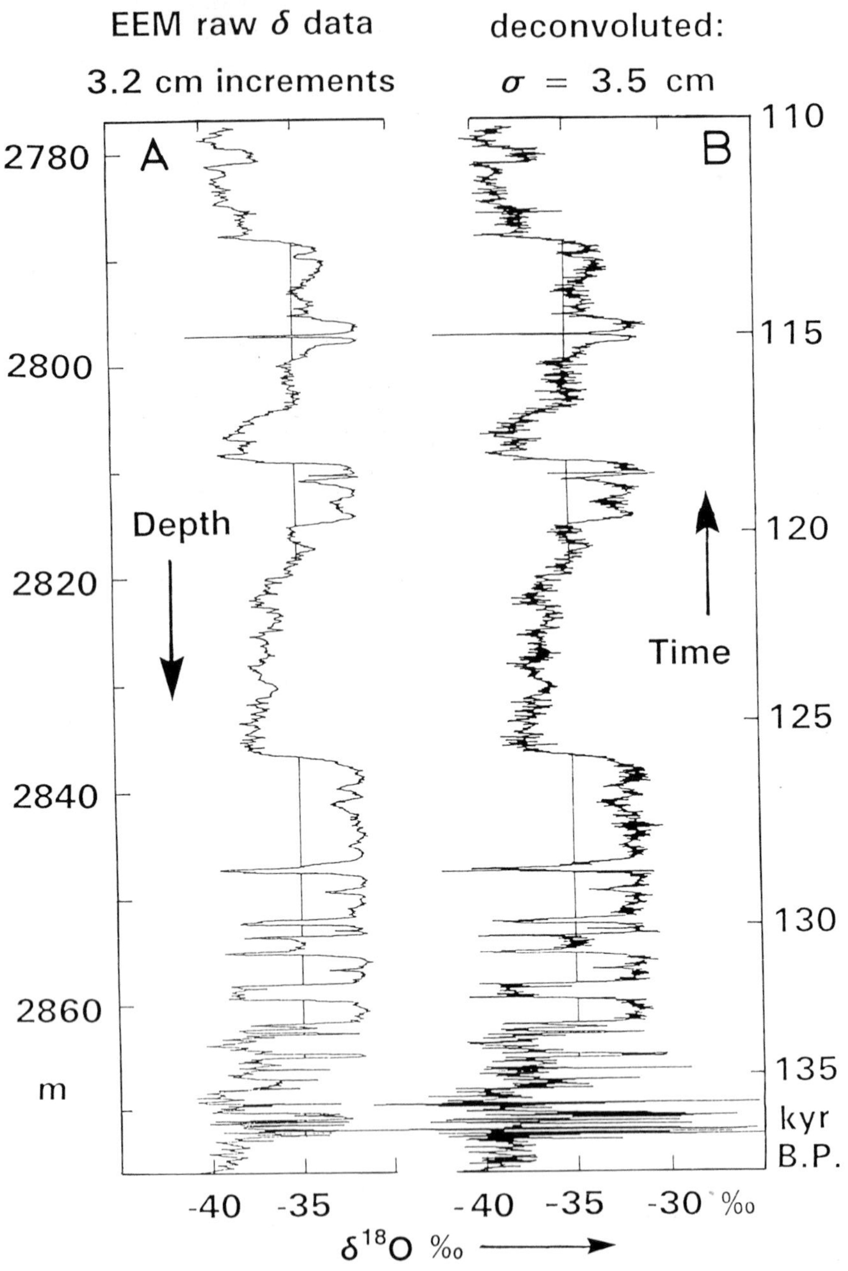

Fig. 4   The Eemian δ record in the GRIP ice core from Summit. A: $\delta^{18}O$ profile measured on 3.4 cm samples spanning the depth interval 2777 to 2880 m. B: Deconvolution of A. Essential disturbances appear only in the depth interval 2869 to 2873 m. The vertical lines show the present δ value

Large-scale folding caused by bedrock obstacles hardly exists at Summit, however, because (1) the bedrock is gently sloping ($\leq$ 40 m per km) in a large area around the drill site, according to radio-echo soundings (HODGE et al., 1990) and gravity measurements (Tscherning, pers. comm.); (2) the shape of internal radar reflection layers (L. Hempel, pers. comm.) suggests that the long-term position of the ice divide was only 5 km west of the present Summit, so the ice movement at Summit seems to have been essentially vertical in the past, confirming that the ice has not travelled long distances over hilly bedrock; (3) folding would imply that the drill penetrated the same layer more than once; it is not possible, however, to find two low δ layers of identical isotopic and chemical composition, at least down to a depth of 2870 m, corresponding to 136 kyr B.P. in our time scale (C. U. Hammer and K. Fuhrer, pers. comm.); and (4) visible cloudy bands[2] indicative of former snow surfaces lie essentially perpendicular to the core axis (i.e. horizontal) down to a depth of 2847 m, corresponding to 129 kyr B.P.; from this depth, the cloudy bands begin sloping more and more to a maximum of 21°, and from 2900 m (~160 kyr B.P.) follows a 54 m depth interval of apparently disturbed stratigraphy (S. Kipfstuhl, pers. comm.), possibly caused by the boudinage effect; the regular layer sequence is re-established from 2954 m (210 kyr B.P.) down to the top of the silty ice.

At great depths, however, the boudinage effect may have caused small-scale disturbances. Some layers may have thickened at the expense of others now missing in the core. If so, the broad outline of the time scale may still be valid, but the layer sequence will not be strictly continuous, and the δ gradients in the ice will be enhanced beyond the values described by the flow model. This has been checked by deconvolution (JOHNSEN, 1977) of a detailed δ profile (Fig. 4A) using the expected diffusion length σ = 3.5 cm. The resulting curve (Fig. 4B) describes the shape of the δ profile freed from the diffusive smoothing since the time of formation of the ice, except in depth intervals where the ice has been exposed to other gradient increasing processes than the thinning of the layers described by the flow model, e.g. boudinage. In such intervals the deconvolution will result in extreme fluctuations. According to Fig. 4B, such features only occur from 2869 to 2873 m, corresponding to the time interval 136 to 138 kyr B.P. Consequently, we adopt the time scale down to a depth of 2981 m ($\geq$ 48 m above bedrock), corresponding to 250 kyr. B.P., but caution should be applied prior to 129 kyr B.P., and particularly prior to 160 kyr B.P.

6. Discussion

Figure 5 is a composite of four different chronological records. Figure 5C shows the upper 2982 m of the Summit δ record plotted in 200 yr increments on the linear time scale. The

---

[2] The cloudy bands are particularly visible in low δ ice. They consist of micro-bubbles formed around micro-particles, which are deposited particularly in early summer (HAMMER et al., 1978).

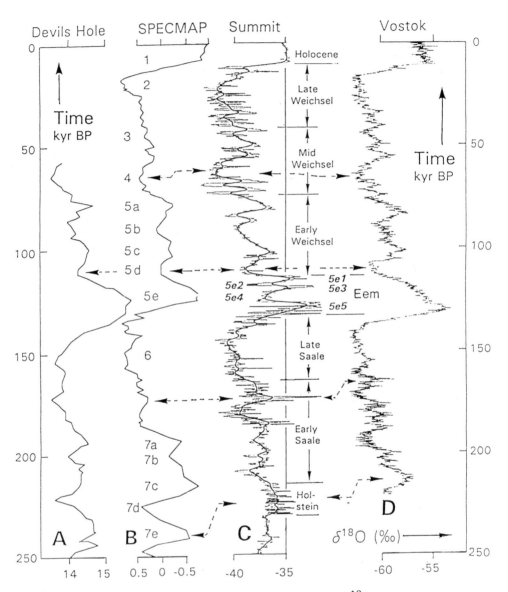

Fig. 5 Four climate records spanning the last two glacial cycles. A: $\delta^{18}O$ variations in vein calcite from Devils Hole, Nevada (Winograd et al., 1992). Dating by U/Th. B: The SPECMAP standard isotope curve (Martinson et al., 1987) with conventional marine isotope stages and sub-stages. Dating by orbital tuning. C: $\delta^{18}O$ record along the upper 2982 m of the GRIP Summit ice core. Each point represents a 200 yr mean value. The smoothed curve is derived by a 5 kyr Gaussian low pass filter. Dating by counting annual layers back to 14.5 kyr B.P., and beyond that by ice-flow modelling. D: δD record from Vostok, east Antarctica (Jouzel et al., 1993), converted into a $\delta^{18}O$ record by the equation $\delta D = 8 \cdot \delta^{18}O + 10‰$. Dating mainly by ice-flow modelling

adopted time scale is further supported, firstly by exhibiting numerous common features (particularly during the last glaciation) in the smoothed version and the other three records in Fig. 5 (a few are indicated by arrows); secondly, because a maximum entropy spectrum (ULRICH & BISHOP; 1975) comprising the total Summit record contains significant signals on the two Milankovitch cycles 41 kyr (obliquity) and, considerably weaker, 24/18 kyr (precession).

The violent δ shifts observed in Greenland cores are less pronounced in the δ record along the Vostok, east Antarctica, ice core (Fig. 5D; JOUZEL et al., 1993), probably because the shifts in Greenland are connected to rapid ocean/atmosphere circulation changes in the North Atlantic region (OESCHGER et al., 1984; BROECKER et al., 1985; DANSGAARD et al., 1989). Both records, as well as the δ record from Devils Hole (Fig. 5A; WINOGRAD et al., 1992), imply MIS 5e (Eem) as an interglacial of considerably longer duration than esti- mated from sea sediment δ records, cp. the SPECMAP record in Fig. 5B (MARTINSON et al., 1987). The disagreement may be explained, in part, by climate instability in the early stages of Eem prolonging the melting of the Saalean ice sheets in America and Eurasia. Further evidence of delayed sea level rise is found in records of the isotopic composition of atmospheric oxygen in the Vostok ice core (JOUZEL et al., 1993). If MIS 5e is interpreted as the period between the first and last years of higher than Holocene δ's, it lasted nearly 20 kyr, from 133 to 114 kyr B.P, according to the Summit record. Another possibility is that 5e began 130 kyr B.P. and ended in the dramatic cold spell 116 kyr B.P., which may be identified with the climatic deterioration that wiped out all trees in northwest Europe, ac- cording to the Grande Pile, France, pollen record (WOILLARD, 1978; DE BEAULIEU & REILLE, 1992).

In Fig. 5C (and in other Summit records as well; GRIP members, 1993), 5e stands out as an interglacial abruptly interrupted several times by periods as cool as 5a and 5c, assuming that the climatic interpretation holds. The duration of 5e2 and 5e4 was apparently 2 and 6 kyr, and they define a tripartition of 5e. CO_2 analyses along the Vostok ice core also suggest a partition of 5e (REYNAUD et al., 1993), but Fig. 5C differs remarkably from most deep sea and European pollen records, which show 5e as a generally warm and stable period, cf. e.g. Fig. 5B and MENKE & TYNNI (1984).

As regards sea sediment δ records, they are primarily indicative of the continental ice vol- ume, which does not necessarily vary with temperature during times of no ice in North America and Eurasia. Furthermore, the resolution is usually limited by bioturbation and coarse sampling.

It is more difficult to reconcile a strongly varying Eemian climate in Greenland with a simultaneous warm and stable European climate as indicated by the pollen records from western Europe (MENKE & TYNNI, 1984). Global climate changes are indeed generally enhanced at higher latitudes, and more pronounced in winter than in summer. The cooling

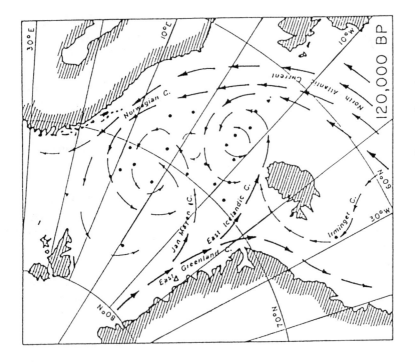

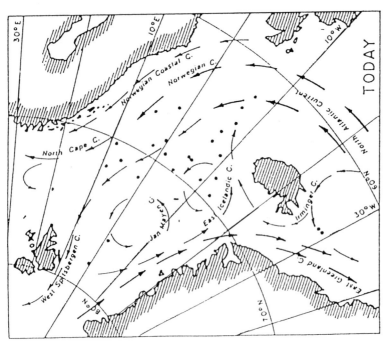

Fig. 6 A. Ocean currents in the Norwegian Sea at persent and B. during marine isotope stage **5e** as derived from studies of ocean sediment cores (from KELLOGG, 1980)

in western Europe corresponding to the long lasting δ minimum **5e4** in Fig. 5C may therefore not have been deep enough during the growing season to cause drastic and visible changes in the European vegetation records. However, indicator plants like holly (*Ilex*) and ivy (*Hedera*), which require relatively warm winter temperatures, occurred frequently throughout the entire Eemian period in western Europe and seem to exclude prolonged intervals of winter cold. Periods colder than now in Greenland and simultaneously warmer than presently in Europe are thus strongly suggested.

This might be possible if, during these periods, the "westerlies" at mid-to-high latitudes in the North Atlantic region had a generally stronger northward component in response to the increased longitudinal temperature gradient across the Norwegian Sea. This would enhance the intensity of the warm Gulf Stream as well as the cold East Greenland Current and give rise to increased deep water formation. Such a circulation pattern is supported by analyses of ocean sediments (KELLOGG, 1980), which suggest (1) that during Eemian time the Gulf Stream was displaced toward east, running closer to the Norwegian coast than now; (2) that the permanent sea ice cover along the east Greenland coast often reached farther south than now; and (3) the cold East Greenland Current broadened to cover most of the Norwegian Sea with a southward branch running east of Iceland (Fig. 6). Alternations between this flow pattern and one similar to that of today would seem to create climate instability in Greenland and simultaneous stability in Europe.

Since the ESF conference in April 1993, the American GISP2 core drilling to bedrock has been completed 30 km west of Summit. The $\delta^{18}O$ profile is in accordance with the GRIP record back to 110,000 yr B.P., but the two records are different in older strata (GROOTES et al. (1993), Nature 366, 552-554), which are disturbed in the GISP2 core. Whether or not this is also the case in the GRIP core is still being discussed.

Acknowledgements

This work is a contribution to the Greenland Ice-Core Project (GRIP) organized by the European Science Foundation. We thank the GRIP participants and supporters for their cooperative effort. We also thank the National Science Foundations in Belgium, Denmark, France, Germany, Iceland, Italy, Switzerland and the United Kingdom, as well as the XII Directorate General of CEC, the Carlsberg Foundation and the Commission for Scientific Research in Greenland for financial support.

References

BARD, E.; ARNOLD, M.; FAIRBANKS, R. G. & HAMELIN, B. (1993): ^{230}Th - ^{234}U and ^{14}C ages obtained by mass spectrometry on corals. Radiocarbon 35, 191-199

BEHRE, K.-E. & VAN DER PLICHT, P. (1992): Towards an absolute chronology for the last glacial period in Europe: radiocarbon dates from Oerl, northern Germany. Vegetation History and Archaeobotany 1, 111-117

BEAULIEU, J.-L. DE & REILLE, M. (1992): The last climatic cycle at la Grande Pile (Vosges, France). A new pollen profile. Quat. Sci. Rev. 11, 431-438

BROECKER, W. S.; PETEET, D. M. & RIND, D. (1985): Does the ocean-atmosphere system have more than one stable mode of operation? Nature 315, 21-26

DAHL-JENSEN, D. & JOHNSEN, S. J. (1986): Palaeotemperatures still exist in the Greenland ice sheet. Nature 320, 250-252

DANSGAARD, W. & JOHNSEN, S. J. (1969): A flow model and a time scale for the ice core from Camp Century, Greenland. J. Glaciol. 53, 215-223

DANSGAARD, W.; JOHNSEN, S. J.; CLAUSEN, H. B. & GUNDESTRUP, N. (1973): Stable isotope glaciology. Medd. Grønl. 197/2, 53 p.

DANSGAARD, W.; CLAUSEN, H. B.; GUNDESTRUP, N.; HAMMER, C. U.; JOHNSEN, S. J.; KRISTINSDOTTIR, P. M. & REEH, N. (1982): A new Greenland deep ice core. Science 218, 1273-1277

DANSGAARD, W.; JOHNSEN, S. J. & WHITE, J. C. (1989): The abrupt termination of the Younger Dryas climate event. Nature 339, 532-534

DANSGAARD, W.; JOHNSEN, S. J.; CLAUSEN, H. B.; DAHL-JENSEN, D.; GUNDESTRUP, N. S.; HAMMER, C. U.; HVIDBERG, C. S.; STEFFENSEN, J. P.; SVEINBJÖRNSDOTTIR, A. E.; JOUZEL, J. & BOND, G. (1993): Evidence for general instability of past climate from a 250-kyr ice-core record. Nature 364, 218-220

GRIP members (1993): Climate instability during the last interglacial period recorded in the GRIP ice core. Nature 364, 203-207

GUNDESTRUP, N. & JOHNSEN, S. J. (1985): A battery powered, instrumented deep ice core drill for liquid filled holes. In: Langway, C. C. Jr.; Oeschger, H. & Dansgaard, W. (eds.): Greenland Ice Cores: Geophysics, Geochemistry and Environment. American Geophysical Union (AGU) Monograph 33, 19-22

HAMMER, C. U.; CLAUSEN, H. B.; DANSGARD, W.; GUNDESTRUP, N.; JOHNSEN, S. J. & REEH, N. (1978): Dating of Greenland ice cores by flow models, isotopes, volcanic debris, and continental dust. J. Glaciol. 20, 3-26

HAMMER, C. U. (1980): Acidity of polar ice cores in relation to absolute dating, past volcanism, and radio-echoes. J. Glaciol. 25, 359-372

HAMMER, C. U.; CLAUSEN, H. B.; DANSGAARD, W.; NEFTEL, A.; KRISTINSDOTTIR, P. & JOHNSON, E. (1985): Continuous impurity analysis along the Dye-3 deep core. In: Langway, C. C. Jr.; Oeschger, H. & Dansgaard, W. (eds.): Greenland Ice Cores: Geophysics, Geochemistry and Environment. American Geophysical Union (AGU) Monograph 33, 90-94

HODGE, S. M.; WRIGHT, D.L.; BRADLEY, J.A.; JACOBEL, R.W.; SKOU, N. & VAUGHN, B. (1990): Determination of the surface and bed topography in Central Greenland. J. Glaciol. 36/122, 17-30

JOHNSEN, S. J. (1977): "Stable isotope homogenization of polar firn and ice". Proc. of Symp. on Isotopes and Impurities in Snow and Ice, Int. Ass. of Hydrol. Sci., Commission of Snow and Ice, International Union of Geophysics and Geodesy (IUGG), XVI General Assembly, Grenoble Aug. / Sept., 1975. International Association of Scientific Hydrology (IASH) Publ. 118, 210-219

JOHNSEN, S. J.; DANSGAARD, W. & WHITE, J. W. C. (1989): The origin of Arctic precipitation under present and glacial conditions. Tellus 41B, 452-468

JOHNSEN, S. J. & DANSGAARD, W. (1992): On flow model dating of stable isotope records from Greenland ice cores. Proceedings of the NATO workshop "The Last Deglaciation: Absolute and Radiocarbon Chronologies" held in Erice, Italy, December 10-12, 1991. NATO ASI Series, I 2., 13-24

JOHNSEN, S. J.; CLAUSEN, H. B.; DANSGAARD, W.; FUHRER, K.; GUNDESTRUP, N.; HAMMER, C. U.; IVERSEN, P.; STEFFENSEN, J. P.; JOUZEL, J. & STAUFFER, B. (1992): Irregular glacial interstadials recorded in a new Greenland ice core. Nature 359, 311-313

JOUZEL, J.; BARKOV, N. I.; BARNOLA, J. M.; BENDER, M.; CHAPPELLAZ, J.; GENTHON, J. C.; KOTLYAKOV, V. M.; LIPENKOV, V.; LORIUS, C.; PETIT, J. R.; RAYNAUD, D.; RAISBECK, G.; RITZ, C.; STIEVENARD, M.; YIOU, F. & YIOU, P. (1993): Extending the Vostok ice-core record of palaeoclimate to the penultimate glacial period. Nature 364, 407-412

KELLOG, T. B. (1980): Palaeoclimatology and palaeo-oceanography of the Norwegian and Greenland seas: glacial-interglacial contrasts. Boreas 9, 115-137

LANGWAY, C.C. Jr.; OESCHGER, H. & DANSGAARD, W. (1985): The Greenland Ice Sheet Program in perspective. In: Langway, C. C. Jr.; Oeschger, H. & Dansgaard, W. (eds.): Greenland Ice Cores: Geophysics, Geochemistry and Environment. American Geophysical Union (AGU) Monograph 33, 1-8

MARTINSON, D. G.; PISIAS, N. G.; HAYS, J. D.; IMBRIE, J.; MOORE, T. C. Jr. & SHACKELTON, N. J. (1987): Age dating and the orbital theory of the Ice Ages: Development of a high-resolution 0 to 300,000-year chronostratigraphy. Quat. Res. 27, 1-29

MENKE, B. & TYNNI, R. (1984): Das Eem-Interglazial und das Weichel-Früglazial von Rederstall/Dithmarschen und ihre Bedeutung für die mitteleuropäische Jungpleistozän-Gliederung. Geol. Jahrb. A76, 1-117

OESCHGER, H.; BEER, J.; SIEGENTHALER, U.; STAUFFER, B.; DANSGAARD, W. & LANGWAY, C. C. Jr. (1984): Lateglacial climate history from ice cores. In: Hansen, J. E. & Takahashi, T. (eds.): Climate Processes and Climate Sensitivity. American Geophysical Union (AGU) Monograph 29/5, Maurice Ewing, 299-306

RAYNAUD, D.; JOUZEL, J.; BARNOLA, J. M.; CHAPPELLAZ, J.; DELMAS, R. J. & LORIUS, C. (1993): The ice record of greenhouse gases. Science 259, 926-934

STAFFELBACH, T.; STAUFFER, B. & OESCHGER, H. (1988): A detailed analysis of the rapid changes in ice-core parameters during the last Ice Age. Ann. Glaciol. 10, 167-170

UEDA, H. T. & GARFIELD, D. E. (1968): Drilling through the Greenland Ice Sheet. U.S. Army Corps of Engineers, Cold Regions Research and Engineering Laboratory, Hanover N.H., Special Report 126, 7 p.

ULRICH, T. J. & BISHOP, T. N. (1975): Maximum entropy spectral analysis and autoregressive decomposition. Rev. Geophys. 13, 183-200

Winograd, I. J; Coplen, T. B.; Landwehr, J. M.; Riggs, A. C.; Ludwig, K. R.; Szabo, B. J.; Kolesar, P. T. & Revesz, K. M. (1992): Continuous 500,000-year climate record from vein calcite in Devils Hole, Nevada. Science 258, 255-260

Woillard, G.M. (1978): Grande Pile peat bog: A continuous pollen record for the last 140,000 years. Quat. Res. 9, 1-21

Authors' addresses:

Drs. W. Dansgaard, S. J. Johnsen, H. B. Clausen, N. Gundestrup, C. U. Hammer & H. Tauber, The Niels Bohr Institute of Physics, Astronomy and Geophysics, Department of Geophysics, University of Copenhagen, Haraldsgade 6, DK-2200 Copenhagen

Solar and climatic components of the atmospheric ^{14}C record

Minze Stuiver

Summary

Palaeo-atmospheric $^{14}CO_2$ varies with changes in ^{14}C production in the atmosphere, or $^{14}CO_2$ transfer between the various earth carbon reservoirs. The former influence affects the Holocene $^{14}CO_2$ record with solar and geomagnetic modulation of the cosmic ray flux. The latter causes climate related re-distribution of $^{14}CO_2$. Because oceanic change plays an important role in both climate and the $^{14}CO_2$ record, similar trends in climate and palaeo-$^{14}CO_2$ do not always prove a solar origin for climatic change.

Zusammenfassung

Der atmosphärische $^{14}CO_2$-Pegel in der Vergangenheit wird sowohl durch Änderungen in der ^{14}C-Produktion als auch im Austausch zwischen den verschiedenen Reservoiren der Erde bestimmt. Der erstere Einfluß wirkt über die Modulation der kosmischen Höhen-strahlung durch die Sonne und das Erdmagnetfeld auf den holozänen ^{14}C-Verlauf, während der letztere eine vom Klima abhängige Umverteilung von $^{14}CO_2$ zur Folge hat. Ähnliche Trends im Verlauf von Klima und $^{14}CO_2$-Pegel allein belegen nicht immer einen gemein-samen solaren Ursprung, weil auch der Ozean selbst sowohl das Klima als auch den $^{14}CO_2$-Pegel beeinflußt.

1. Introduction

Palaeo-records of ^{14}C, ^{10}Be and other cosmogenic isotopes contain important solar change information. The solar wind's magnetic (helio-magnetic) modulation of the cosmic ray flux causes substantial variability in production rates of cosmogenic isotopes in the earth's atmosphere and surface crust and thus leads to concentration changes of these isotopes. Given a suitable recording medium a history (proxy record) of cosmogenic isotope con-centration change can usually be derived. Such proxy records may be obscured as isotope concentration can also be effected by other, non-solar, causes. The helio-magnetic signal has to be separated from such background "noise".

The environmental recorders of choice have been, so far, ice, trees and corals. Trees, with their annual ring deposition, yield a detailed record of cosmogenic ^{14}C concentration whereas polar ice, whenever annually layered, provides similar cosmogenic ^{10}Be concen-tration information. However, proxy records and instrumental records are not synonymous.

Whereas our selection of instruments minimizes distortion and optimizes signal to noise ratios the (palaeo)environmental recorders have not been optimized by an experimenter. The environmental recorder itself distorts and adds noise. For instance, when depositing cellulose as a photosynthetic endproduct, the tree may erratically discriminate against the heavier carbon isotopes (^{13}C and ^{14}C). Such discrimination usually is given as a $\delta^{13}C$ value, with $\delta^{13}C$ equal to the per mill (or percent) deviation of the sample's $^{13}C/^{12}C$ ratio from that of a standard. The tree's discrimination influences the ^{14}C concentration record in a similar manner. Fortunately ^{13}C and ^{14}C are both subjected to this type of biological discrimination and the relationship between ^{13}C and ^{14}C isotope discrimination (^{14}C fractionation is twice that of ^{13}C) can be applied to normalize the ^{14}C record on a fixed ^{13}C isotope ratio ($\delta^{13}C$ = -25‰ relative to the PDB standard). The correction removes from the ^{14}C record biological noise inherent to the photosynthetic cycle, but such removal is only exact if the sole determinant of the ^{13}C record is photosynthetic change.

$\Delta^{14}C$ values, which are the per mill (or percent) deviations of the sample's $^{14}C/^{12}C$ ratio from that of a standard (usually National Bureau of Standards oxalic acid), always include $\delta^{13}C$ normalization. Sample age factors into tree-ring $\Delta^{14}C$ calculations because the sample's $^{14}C/^{12}C$ ratio has to be back calculated from the measured ratio to take radioactive decay into account.

Whereas heliomagnetic modulation of the cosmic ray flux results in cosmogenic production rate change, the environmental recorders register concentration change. The transformation of production rate change into concentration change can be rather complicated. For steady state ^{14}C conditions (here the radioactive decay rate equals the production rate) there are a total of 8240 years of ^{14}C production in the various earth carbon reservoirs. Approximately 100 years of ^{14}C production are in the atmosphere (CO_2) and 300 years in the active biosphere (organic materials), with nearly the entire remainder in the oceans (dissolved CO_2, bi-carbonate and carbonate ion) and sediments. For an atmosphere independent of the other reservoirs a ^{14}C production rate increase of, for instance, 25% from one year to the next increases atmospheric ^{14}C concentration by 1/4 yr per 100 yr, or 2.5‰. The actual atmospheric change will be even less as carbon (and ^{14}C) actively exchanges between the atmosphere and the biosphere as well as the oceans. Thus a single year production rate signal is attenuated at least 100 times. Given sufficient time, a signal lasting many millennia will not be attenuated as a 25% higher steady state ^{14}C concentration will ultimately be reached. The signal attenuation (and phase shift), which is very much time dependent, complicates the conversion of concentration change into production rate change (Q) and the investigator has to resort to complicated carbon reservoir calculations (carbon reservoir "modeling") that include approximate size, and exchange rate, of earth's various carbon reservoirs.

Not all cosmogenic isotopes show the drastic transformation of the production rate signal found for ^{14}C. The ^{10}Be concentration (e.g. in polar ice) reflects ^{10}Be production rather closely as this cosmogenic isotope has a relatively short residence time in the atmosphere (stratosphere-troposphere exchange is in the 1 to 2 year range). Once in the troposphere, removal on cloud condensation nuclei can be a matter of weeks. Quite opposite to ¹⁴C, newly produced ¹⁰Be does not accumulate over several years in the troposphere and the signal attenuation related to dilution in this reservoir will be relatively minor. Thus, whereas on a decade scale we find solar related Δ¹⁴C changes of the order of 2 per mill, the corresponding ¹⁰Be concentration changes in atmospheric precipitation in polar regions are more like 20%. The fast removal of ¹⁰Be is a mixed blessing, however, because ¹⁰Be in precipitation is not tropospherically averaged like ¹⁴C. ¹⁰Be concentrations depend on geomagnetic latitude and substantial changes in local ¹⁰Be concentration records may occur when atmospheric circulation shifts into another mode.

The latitude dependence of cosmogenic isotope production is the result of the interaction of the earth's magnetic field with the incoming (electrically charged) cosmic ray particles whereas altitude dependence (causing a production maximum in the upper atmosphere) is tied to the level of interaction of cosmic ray particles with the atmospheric nuclei. The sum total of these dependencies on cosmogenic isotope production is depicted in Fig. 1 where calculated cosmogenic ¹⁴C isotope production rates are given for a twentieth century geomagnetic dipole configuration. ¹⁰Be palaeoconcentrations are preferably, and of necessity, measured in polar regions where solar minimum-solar maximum differences are largest.

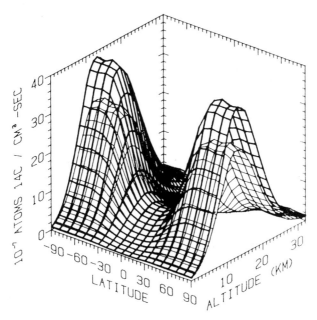

Fig. 1 ¹⁴C production rates in the atmosphere during the 1958 sunspot maximum (light "lower" lines) and 1954 sunspot minimum (heavy "upper" lines). Based on calculations of T. F. Braziunas of the Quaternary Isotope Laboratory

Besides modification inherent to signal transformation and recorder noisiness the ^{14}C proxy record also will reflect climate and geomagnetic dipole change. Geomagnetic dipole moment (M) change influences production rate (inversely proportional to the square root of M). The total amount of cosmogenic ^{14}C in the earth's exchangeable carbon reservoirs is solely defined by the "external" forcing factors (geomagnetic dipole intensity and solar wind magnetic properties) but what is actually measured in samples is the ^{14}C/^{12}C ratio of carbon of specific reservoirs (e.g. the atmosphere or the ocean). Since climate related changes in air-sea exchange rate and other physical properties (such as ocean circulation) can influence the ^{14}C distribution among the reservoirs, the atmospheric ^{14}C/^{12}C ratio does not only depend on cosmic ray flux change but also on mechanisms that cause ^{14}C re-distribution. Changes that can be tied to climate are thermohaline circulation ("conveyer belt") change of the oceans and wind speed changes that influence the CO_2 exchange between the oceans and the atmosphere.

2. Atmospheric Δ^{14}C during the Holocene

The atmospheric Δ^{14}C record contains components attributable to (1) solar modulation of the cosmic ray flux, (2) convection currents in the interior of the earth that influence the geomagnetic dipole moment, and (3) variability in thermohaline ocean circulation and atmosphere-ocean CO_2 gas exchange rate (this forcing factor also relates to concurrent climatic change). Separation of these components depends on their presumed relative importance. The separation is also frequency dependent since long-term (millennia or more) Δ^{14}C change usually is attributed to forcing factor (2) and (3). These forcing factors are presumed to be so dominant on this time scale that genuine millennia type solar variability (if any) cannot be unequivocally proven. Removal of the long-term trends (e.g. by removing a 2000-year moving average) leads to residual Δ^{14}C profiles such as the one given in Fig. 2. Solar modulation manifests itself here in the decade to several centuries time domain.

The Fourier power spectrum of the Fig. 2 residual Δ^{14}C record has periods exceeding the 2s significance level at 512 yr, 206 yr, 148 yr, 87 yr and 46-49 yr (STUIVER & BRAZIUNAS, 1993), with the spectral power of the 512-year cycle concentrated in the earliest portion of the Δ^{14}C record. This cycle has been attributed to oceanic forcing by the above authors. The postulated mechanism for oceanic Δ^{14}C change is *via* the (unstable) salt oscillator in the North Atlantic where salinity lowering results in increased stratification and reduced deep water formation. Reduced deep water formation in turn parallels reduced upwelling of ^{14}C deficient water elsewhere in the oceans thus (temporarily) increasing surface ocean and atmospheric Δ^{14}C levels.

Although the possibility of an incremental oceanic contribution cannot be dismissed out of hand, solar modulation of the cosmic ray flux must play a crucial role in the Δ^{14}C variance associated with the 206, 148 and 87-year periodicities (e.g. STUIVER & QUAY, 1980;

STUIVER & BRAZIUNAS, 1989). The dependence of cosmogenic ^{14}C production on solar wind properties (with increased ^{14}C production tied to lower sunspot numbers) is well established for the 11-year sunspot cycle (LINGENFELTER, 1963; O'BRIEN, 1979). A production rate change of longer duration evidently contributed to the late seventeenth century Δ^{14}C maximum of Fig. 2. Apart for a small time lag, this Maunder Δ^{14}C maximum is contemporary with the A.D. 1650-1715 Maunder minimum (EDDY, 1976) in sunspot numbers. By using the twentieth century observed relationship between the 11-year cycles in sunspot numbers and atmospheric neutron production rate (which in turn is proportional to ^{14}C production rate) the changes in ^{14}C production rate during the Maunder Minimum can be calculated. This calculation leads to a Maunder Δ^{14}C increase of about half the observed one when assigning zero sunspot numbers to the entire 65-year interval (STUIVER & QUAY, 1980).

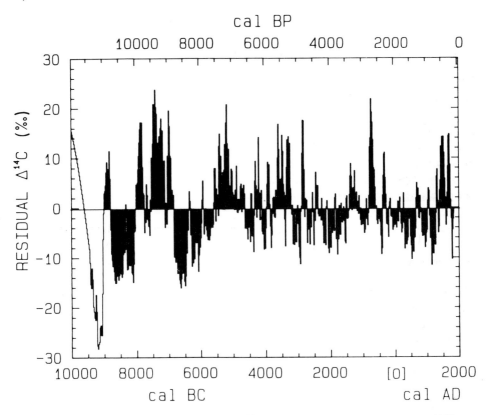

Fig. 2 Residual Δ^{14}C obtained by deducting a Δ^{14}C long-term trend approximating a 2000-year moving average. The oldest, non-shaded oscillation is partly (first 500 calibrated (cal) years) derived from coral determinations (BARD et al., 1993). Remaining Δ^{14}C values were derived from tree-ring measurements (STUIVER & REIMER (1993), from data of KROMER & BECKER (1993), LINICK et al (1986), PEARSON & STUIVER (1993) and STUIVER & PEARSON (1993)). The seventeenth century Δ^{14}C maximum is associated with the Maunder sunspot minimum

The Maunder $\Delta^{14}C$ amplitude discrepancy can be removed, however, when postulated long-term trends in solar modulation are taken into account (STUIVER & QUAY, 1980). Solar modulation considerations of LAL (1992), however, would limit the maximum ^{14}C production rate change during the Maunder Minimum to about 2/3 the increase needed to generate the observed $\Delta^{14}C$ augmentation in carbon reservoir models. Evidently solar modulation of the cosmic ray flux explains at least one half of the Maunder $\Delta^{14}C$ increase. The remaining half can either be attributed to additional cosmic ray flux increases when sunspots are absent (STUIVER & QUAY, 1980), or to oceanic influences. For the latter scenario the solar signal would have been augmented by a synchronous oceanic response.

Thus, if climate related ocean circulation change plays a role, it must have enhanced the Maunder $\Delta^{14}C$ signal. A postulated mechanism is reduced upwelling (and deep water formation) during episodes of climatic change induced by lower solar activity. This mechanism has been explored by STUIVER & BRAZIUNAS (1993) as an alternative to the 100% solar origin for the Maunder $\Delta^{14}C$ maximum. This modification entails a minor solar induced Maunder Minimum climate change amplified by salinity effects on North Atlantic thermohaline circulation.

3. The single year $\Delta^{14}C$ component

A nearly 450-year (A.D. 1510-1954) $\Delta^{14}C$ annual time series, described by STUIVER & BRAZIUNAS (1993), contains spectral power near the 56, 26, 17, 13, 10.4, 6.4, 5.7, 3.5 and 2.7-2.1-year periodicities. The "11-year " sunspot cycle relates to the 10.4-year periodicity, and perhaps to some of the longer periods. Agreement between inverted $\Delta^{14}C$ and sunspot numbers is evident for several individual 11-year cycles, and, even though the low sunspot numbers associated with the Maunder Minimum suggest minimal 11-year modulation, the magnitudes of the 10.4-year $\Delta^{14}C$ oscillations during this interval do not differ appreciably from those observed for the preceding and following years (STUIVER & BRAZIUNAS, 1993). The 11-year production rate also can be seen in the ^{10}Be concentrations of polar ice (BEER et al., 1990), but with correlation coefficients between calculated ^{14}C production rates and ^{10}Be concentration limited to 0.42 for the post-Maunder era. Evidently other variables, such as climate, play a major role in either ^{10}Be or ^{14}C variance, or both.

Climatic influences on the $\Delta^{14}C$ record include variable rates of upwelling associated with the El Niño-Southern Oscillation (ENSO) events as well as North Atlantic thermohaline circulation change. The 2 to 6.4-year periodicities in the $\Delta^{14}C$ record correspond in duration to the irregular cycle of ENSO events (QUINN & NEAL, 1992) whereas irregular thermohaline oscillations of approximately 40 to 50 years were noted by DELWORTH et al. (1993) in a coupled ocean atmosphere model.

4. The search for a sun-climate connection

Atmospheric Δ^{14}C values represent globally integrated ^{14}C/^{12}C ratios. A search for a Δ^{14}C-climate relationship is basically an attempt to match global Δ^{14}C with global climate. However, global climate coverage of trends, derived from instrumental records, is only available for the past few hundred years. Beyond this time span, only regional proxy data are available and the search for a Δ^{14}C-climate connection usually involves the comparison of regional climate records to the global Δ^{14}C tracer. Several studies (e.g. SONETT & SUESS, 1984; WIGLEY & KELLY, 1990; ANDERSON, 1992; TAYLOR et al., 1992; MAGNY, 1993; COOK et al., 1992, 1994; SCUDERI, 1993) have suggested a relationship between regional climate and Δ^{14}C but the relatively small correlation coefficients (less than 0.2) imply that only a few percent (if any) of the regional climatic variance can be tied to global Δ^{14}C variance.

A relationship between climate and Δ^{14}C would be a matter of course if global Δ^{14}C perturbations of the century variety had been caused by oceanic thermohaline circulation-climate change. As the North Atlantic is the prime candidate for thermohaline circulation and related Δ^{14}C change, climate proxy data "downwind" from this region (Europe) would relate to global Δ^{14}C to a larger extent than climate proxies from elsewhere. But the Δ^{14}C record, in addition to an oceanic component, also contains substantial solar mediated cosmic ray production rate variance. Here a Δ^{14}C-climate connection could be interpreted as a solar induced coeval climatic and Δ^{14}C change.

Solar and oceanic components of the Δ^{14}C record are, at least to some extent, concentrated in different frequency domains (e.g. the 512-year oceanic versus the 206-year solar periodicity). Attempts to demonstrate a solar influence on climate through Δ^{14}C-climate comparisons have, so far, not taken into account the complexity of the Δ^{14}C signal. A convincing demonstration of the solar influence on climate by way of Δ^{14}C is therefore not yet available.

Acknowledgements

The U.S. National Science Foundation supported the ^{14}C work discussed here. P. J. Reimer and P. Wilkinson provided technical help and T. F. Braziunas provided the Figure 1 calculations in addition to valuable suggestions for improvement of the manuscript.

References

ANDERSON, R. Y. (1992): Possible connection between surface winds, solar activity and the earth's magnetic field. Nature 358, 51-53

BARD, E.; ARNOLD, M.; FAIRBANKS, R. G. & HAMELIN, B. (1993): ^{230}Th-^{234}U and ^{14}C ages obtained by mass spectrometry on corals. Radiocarbon 35, 191-199

BEER, J.; BLINOV, A.; BONANI, G.; FINKEL, R. C.; HOFMANN, H. J.; LEHMANN, B.; OESCHGER, H.; SIGG, A.; SCHWANDER, J.; STAFFELBACH, T.; STAUFFER, B.; SUTER, M. & WÖLFLI, W. (1990): Use of ^{10}Be in polar ice to trace the 11-year cycle of solar activity. Nature 347, 164-166

COOK, E. R.; BIRD, T.; PETERSON, M.; BARBETTI, M.; BUCKLEY, B. M.; D'ARRIGO, R. & FRANCEY, R. (1992): Climatic change over the last millennium in Tasmania reconstructed from tree-rings. The Holocene 2, 205-217

COOK, E. R.; BUCKLEY, B. M. & D'ARRIGO, R. D. (1994): Decadal-scale oscillatory modes in a millennia-long temperature reconstruction from Tasmania. Proc. Nat. Acad. Sci. Workshop on "The natural variability of the climate system on 10-100 year time scales", Irvine, Cal. Sept. 21-24, 1992

DELWORTH, T.; MANABE, S. & STOUFFER, R. J. (1993): Interdecadal variations of the thermohaline circulation in a coupled ocean-atmosphere model. J. Climate 6, 1993-2011

EDDY, J. A. (1976): The Maunder Minimum. Science 192, 1189-1202

KROMER, B. & BECKER, B. (1993): German oak and pine ^{14}C calibration, 7200-9400 B.C. Radiocarbon 35, 125-135

LAL, D. (1992): Expected secular variations in the global terrestrial production of radiocarbon. In: Bard, E. & Broecker, W. S. (eds.): The last deglaciation: absolute and radiocarbon chronologies. NATO ASI Series 12, Springer Verlag Heidelberg, 113-126

LINGENFELTER, R. E. (1963): Production of carbon-14 by cosmic-ray neutrons. Rev. Geophys. 1, 35-55

LINICK, W. L.; LONG, A.; DAMON, P. E. & FERGUSON, C. W. (1986): High-precision radiocarbon dating of Bristlecone pine from 6554 to 5350 B.C. Radiocarbon 28, 943-953

MAGNY, M. (1993): Solar and geomagnetic influences on Holocene climatic changes. Quat. Res. 40, 1-9

O'BRIEN, K. (1979): Secular variations in the production of cosmogenic isotopes. J. Geophys. Res. 84, 423-431

PEARSON, G. W. & STUIVER, M. (1993): High-precision bi-decadal calibration of the radiocarbon time scale, 500-2500 B.C. Radiocarbon 35, 25-33

QUINN, W. H. & NEAL, V. T. (1992): The historical record of El Niño events. In: Bradley, R. S. & Jones, P. D. (eds.): Climate since A.D. 1500. Routledge, Chapmann and Hall, 623-648

SCUDERI, L. A. (1993): A 2000-year tree-ring record of annual temperatures in the Sierra Nevada Mountains. Science 259, 1433-1436

SONETT, C. P. & SUESS, H. E. (1984): Correlation of Bristlecone pine ring widths with atmospheric ^{14}C variations: a climate-sun relation. Nature 307, 141-143

STUIVER, M. & BRAZIUNAS, T. F. (1989): Atmospheric ^{14}C and century-scale solar oscillations. Nature 338, 405-408

STUIVER, M. & BRAZIUNAS, T. F. (1993): Sun, ocean, climate and atmospheric ^{14}CO$_2$: an evaluation of causal and spectral relationships. The Holocene 3, 289-304

STUIVER, M. & PEARSON, G. W. (1993): High-precision bidecadal calibration of the radio-carbon time scale, A.D. 1950-500 B.C. and 2500-6000 B.C. Radiocarbon 35, 1-23

STUIVER, M. & QUAY, P. D. (1980): Changes in atmospheric carbon-14 attributed to a variable sun. Science 207, 11-19

STUIVER, M. & REIMER, P. J. (1993): Extended ^{14}C data base and revised CALIB 3.0 ^{14}C age calibration program. Radiocarbon 35, 215-230

TAYLOR, K.; ROSE, M. & LAMOREY, G. (1992): Relationship of solar activity and climatic oscillations on the Colorado plateau. J. Geophys. Res. 97, 15803-15811

WIGLEY, T. M. L. & KELLY, P. M. (1990): Holocene climatic change, ^{14}C wiggles and variations in solar irradiance. Phil. Trans. Roy. Soc. London A 330, 547-560

Author's address:

Dr. M. Stuiver, Department of Geological Sciences and Quaternary Research Center, University of Washington AK-60, USA-Seattle, WA 98195

Finnish pine tree-ring data as a possible source for solar output research

Matti Eronen & Pentti Zetterberg

Summary[1]

Work is going on in northern Finnish Lapland and the adjacent area of Norway to build a long pine (*Pinus sylvestris*, L.) tree-ring chronology approx. 7000 years in length. So far the continuous ring-width curve extends back to the year 115 B.C., and is the longest "absolutely" dated master chronology for northern Europe. It is highly applicable for palaeoclimatic purposes, as summer temperatures are the dominant factor determining the annual growth of pine in that area (ZETTERBERG et al., in prep.; ERONEN et al., 1994; LINDHOLM et al., in prep.).

There now also exists a long older part at the master chronology, which has been fixed in time by several radiocarbon dates, but at the time of writing this there still is a gap of about 300 years separating the "floating" chronology from the "absolute" one. The older tree-ring series covers about 4500 years extending back to approx. 4900 B.C.

The samples used for building the chronology have been collected mainly from small lakes in the forest limit zone of northern Fennoscandia, with additional material from peat sections and old wooden buildings, and a number of cores from living trees (ERONEN & HUTTUNEN, 1993; ERONEN & ZETTERBERG, in prep.; ERONEN et al., 1994). The ring-width variations measured from living trees are being used for climatic calibration, by comparing them with the available northern weather records (LINDHOLM et al., in prep.).

The discs cut from the subfossil pines are normally 5-15 cm thick, but several larger samples have also been collected in order to provide material for probable further research. These tree-rings are certainly suitable for many kinds of analysis providing information about past climates and environments. The trees are generally well preserved, even though they have died several thousands of years ago. One disadvantage is that the tree-rings are extremely narrow, because trees grow very slowly close to the limit of their distribution. The ring width thickness in the present material (over 1000 subfossil trunks) is most commonly 0.4-0.6 mm, but values higher than 1.5 mm for a single ring occur frequently. The living age of most of the subfossils collected has been 150-250 years (samples containing

---

[1] Since the results presented here have already been published in greater detail (see references), the authors decided only to give a lengthier summary in this volume.

less than 50 annual rings have been discarded in the field), and the oldest tree had reached an age of approx. 520 years.

The living trees take in atmospheric carbon dioxide for photosynthesis and through this lay down a carbon isotope record which can be extracted from the tree-rings. Measurements of the $^{13}C/^{12}C$ ratio in the tree-rings of northern pines are going on (SONNINEN & JUNGNER, in prep.), but it is the ^{14}C values that are of prime interest for research into solar output (STUIVER & BRAZIUNAS, 1989; STUIVER et al., 1991). The Fennoscandian pine tree-ring record constitutes a suitable material for isotopic measurements related to solar output and samples for such purposes can be provided on request.

References

ERONEN, M. & HUTTUNEN, P. (1993): Pine megafossils as indicators of Holocene climatic changes in Fennoscandia. Paläoklimaforschung / Palaeoclimate Research 9, 24-46

ERONEN, M.; LINDHOLM, M. & ZETTERBERG, P. (1994): Extracting palaeoclimatic information from pine tree-rings in Finland for historical climatic proxy data collection. Paläoklimaforschung / Palaeoclimate Research 13, 43-50

ERONEN, M. & ZETTERBERG, P. (in prep.): Expanding dendrochronological data on Holocene changes in the polar/alpine pine limit in northern Fennoscandia. Paläoklimaforschung / Palaeoclimate Research (Skibotn workshop proceedings)

LINDHOLM, M.; ERONEN, M.; MERILÄINEN, J. & ZETTERBERG, P. (in prep.): Climatic calibration of tree-ring width variations in forest-limit pines (Pinus sylvestris, L.) in northern Finnish Lapland. Paläoklimaforschung / Palaeoclimate Research (Skibotn workshop proceedings)

SONNINEN, E. & JUNGNER, H. (in prep.): Stable carbon isotopes in tree-rings of a pine from northern Finland. Paläoklimaforschung / Palaeoclimate Research 15

STUIVER, M. & BRAZIUNAS, T. (1989): Atmospheric ^{14}C and century-scale solar oscillations. Nature 338, 405-407

STUIVER, M.; BRAZIUNAS, T.; BECKER, B. & KROMER, B. (1991): Climatic, solar, oceanic, and geomagnetic influences on Lateglacial and Holocene atmospheric $^{14}C/^{12}C$ change. Quat. Res. 35, 1-24

ZETTERBERG, P.; ERONEN, M. & LINDHOLM, M. (in prep.): The Atlantic/Subboreal climatic shift, tree-ring evidence from northern Fennoscandia. Paläoklimaforschung / Palaeoclimate Research (Skibotn Workshop Proceedings)

Authors' addresses:

Prof. Dr. M. Eronen, Department of Geology, University of Oulu, Linnanmaa, FIN-90570 Oulu
Dr. P. Zetterberg, Karelian Institute, University of Joensuu, P.O. Box 111, FIN-80101 Joensuu

Climatic signatures derived from D/H ratios in the cellulose of late wood in tree-rings from spruce (*Picea abies* L.)

Josef Lipp & Peter Trimborn

Summary

Natural variations in the deuterium content of carbon-bound hydrogen of late wood cellulose from rings of five spruce trees were measured. The δ^2H values in annual late wood of spruce trees (*Picea abies* L.) from Middle Franconia (FRG) show a good correlation with both temperature and relative humidity during July and August (period of late wood formation). This isotope pattern results from the isotopic composition of precipitation and possibly soil water absorbed during (summer) photosynthesis. The δ^2H values of summer precipitation reflect mean July and August temperatures at the growth site, and hence a certain δ^2H cellulose-temperature relationship is reasonable. Furthermore, it is assumed that changes in relative summer humidity may cause changes in δ^2H values of late wood tree-rings due to evapotranspiration effects. A negative relationship between ring width and δ^2H values of late wood tree-rings was observed. This is interpreted as climatic signal, since tree-ring widths reflect climate to some extent, especially in dry and warm, and in cool and moist years ("signature" years). It appears that during the twentieth century climatic oscillations occur in 14-year cycles.

Zusammenfassung

Fünf Fichten (*Picea abies* L.) aus Mittelfranken wurden zwischen 1880 und 1981 bezüglich der Deuteriumgehalte ihrer Jahresringe gemessen. Das Spätholz, welches im Juli und August gebildet wird, läßt sich mit den Juli- und Augusttemperaturen der entsprechenden Jahre korrelieren. Der δ^2H-Wert ist außerdem von der relativen Luftfeuchte dieser Sommermonate abhängig. Die Beziehung zur Temperatur wird dadurch erklärt, daß die Bäume das Niederschlagswasser, dessen δ^2H-Wert temperaturabhängig ist, beim Aufbau der Jahrring-Cellulose verwenden. Die Temperaturabhängigkeit des Deuteriumgehalts des Niederschlags spiegelt sich sozusagen im Deuteriumgehalt der Jahrring-Cellulose wider. Dies wird dadurch verständlich, daß das betrachtete Probengebiet relativ niederschlagsarm ist (ca. 700 mm/Jahr) und die Bäume die seltenen Sommer-Niederschläge sofort durch die Wurzeln aufnehmen müssen. Die Beziehung zur Luftfeuchte beruht auf Verdunstungseffekten, die im Blattwasser zu einer Anreicherung von Deuterium führen. In besonders trockenen, heißen Jahren (Signaturjahren) wird eine Deuterium-Anreicherung in den Jahresringen erkennbar, die mit einer Wachstumsreduktion derselben Ringe einhergeht.

Das Deuteriummuster der Bäume unterliegt klar erkennbaren Rhythmen, die 14-jährige Maxima und Minima zeigen. Diese Rhythmen werden höchstwahrscheinlich durch klimatische Schwankungen (warm, trocken / kühl, feucht) derselben Frequenz hervorgerufen.

1. Introduction

Measurements of the D/H ratios of the non-exchangeable hydrogen in cellulose extracted from tree-rings have shown that climatic information may be obtained from the annual growth layers of trees (YAPP & EPSTEIN, 1982a). In general, D/H ratios in meteoric waters reflect the temperature at the site of precipitation - the lower the temperature the smaller the isotopic ratio and *vice versa* (YURTSEVER & GAT, 1981; DANSGAARD, 1964). The deuterium content of the carbon-bound hydrogen in tree-ring cellulose has been found to be systematically related to the deuterium content of associated meteoric waters (LIBBY & PANDOLFI, 1974; SCHIEGL, 1974; EPSTEIN et al., 1976). Thus, D/H ratios in tree-rings can preserve a record of past temperature variations. Such a climatic "signal" has already been identified for $^{18}O/^{16}O$ ratios of tree cellulose (GRAY & THOMPSON, 1976; BURK & STUIVER, 1981; EPSTEIN et al., 1977; EDWARDS & FRITZ, 1986). With the development of sophisticated analytical techniques the opportunity has arisen for further investigations into the climatic significance of the D/H ratios of carbon-bound hydrogen from tree-ring cellulose (YAPP & EPSTEIN, 1982b). The δ^2H variations in trees offer the possibility that both spatial and temporal climatic signals can be discerned, because trees are geographically widespread and also contain an internal chronology which may isotopically record local climatic change. Studies regarding the climatic significance of δ^2H variations of tree cellulose C-H hydrogen have not been fully satisfactory, largely because of the incapability to separate the climatic δ^2H-signal from non-climatic isotope variations (YAPP & EPSTEIN, 1982a).

The objective of this study was to propose some techniques for extracting the climatic δ^2H signal and thus to evaluate the climatic significance of tree δ^2H values on the temporal scale. D/H ratios were measured on nitrated cellulose in the late wood of single tree-rings from five Franconian spruce trees. It was found that only the late wood, which is generated in summer, reflects climatic information unambiguously (WILSON & GRINSTED, 1977; LIPP et al., 1991; LIPP & TRIMBORN, 1991; NORTHFELT et al., 1981). By correlating D/H ratios with various climatic parameters, the impact of temperature, relative air humidity, and precipitation amount on δ^2H of tree-ring cellulose can be estimated.

2. Materials and methods

We analysed five spruce trees (*Picea abies*) which grew on a 30-40 cm thick clay soil from the Opalinus-Clay-series in a forest near Bad Windsheim, FRG (Schußbach-Wald, 49°30'N,

10°27'E, 330 m a.s.l.). The site has a precipitation rate of about 700 mm per year. In the summer, the upper soil layer is often desiccated, causing cracks in the soil.

The time span of A.D. 1882-1981 was covered by the five trees as follows. Spruce 1: 1907-1981; Spruce 2: 1882-1981; Spruce 3: 1889-1981; Spruce 4: 1882-1981; Spruce 5: 1882-1981. Meteorological data were available at the station Neustadt/Aisch, FRG (49°35'N, 10°37'E, 315 m a.s.l.) for the time from 1950 to 1981. Rainwater was sampled for isotope analysis at the meteorological station of Würzburg, FRG (49°48'N, 9°57'E, 182 m a.s.l.) from 1979 to 1988.

The late wood of the single tree-rings was sampled under the microscope with the aid of a dental drill. Following the Soxhlett extraction of lipids (NORTHFELT et al., 1981), the cellulose was extracted and nitrated (ALEXANDER & MITCHELL, 1949). Nitration is necessary to eliminate the exchangeable hydroxyl hydrogen present in cellulose (EPSTEIN et al., 1976). Cellulose nitrate samples were combusted (NORTHFELT et al., 1981; STUMP & FRAZER, 1973), and the H_2O was reduced to H_2 on hot uranium for D/H analyses. Isotopic data are reported as $\delta(‰)$-values with respect to V-SMOW (Vienna Standard Mean Ocean Water) as an international reference. Repeatability of analytical results (preparation plus mass-spectrometric measurements) was better than $\pm 2‰$ for δ^2H.

3. Results

In Fig. 1, the original (not corrected for trends) δ^2H values of the nitrated late wood cellulose from annual tree-rings of five spruce trees are shown. The aging of spruce trees 1-5 is accompanied by steadily increasing δ^2H values with time. Within about the first 20 years the slope is steep ("juvenile trend"), but it flattens during maturity ("age trend"). The values show large interannual variations, sometimes in the range of about 40‰. During many years the five trees have the same trend in δ^2H. For example, from 1975 to 1976 they all increase, and from 1930 to 1931 they all decrease. Such agreements with all 5 trees are called "100% - signatures", which occur during "signature years". Within the chronology of this study (80 years), 43 signature years occur (54% of the total years). The absolute δ^2H values of the five spruce trees are not identical. The average δ^2H values of the single trees are marked at the y-axis. For example, the average δ^2H values of spruce 5 are lower than those of spruce 2 by 12.2 ‰.

4. Discussion

The difference between juvenile and mature sections of the trees does not coincide with any known climatic event. The answer may be found in the physiological factors controlling tree growth (GRAY & SONG, 1984). There are a number of changes which occur in the

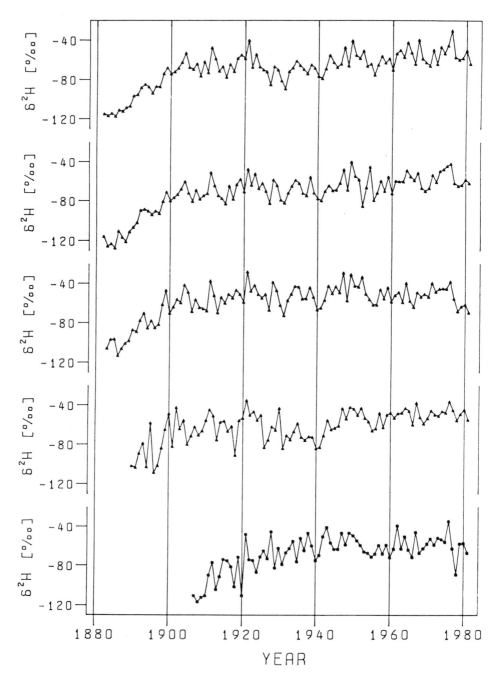

Fig. 1 Original δ²H values of annual late wood tree-ring cellulose from individual spruce trees showing the juvenile and "aging" effect

structure and chemical composition of a cell wall associated with the development from the juvenile stage to that of maturity (DADSWELL & HILLIS, 1962). For example, the average cell length increases through the successive growth rings from the centre of the tree outwards until a more or less constant value is reached (KOZLOWSKI, 1962). There are also changes in the chemical composition as the tree grows, so that, for example, the fraction of cellulose increases through the juvenile growth period (WARDROP, 1951). Thus, as the tree ages the wood formed becomes more uniform in structure and composition. Due to the isotopic competition between chemical constituents in wood, changes in structure and chemical composition may alter the isotopic composition of the wood giving rise to the difference in δ^2H observed in mature and juvenile wood. In dendroclimatology, juvenile rings are often excluded from the analysis because they provide the least reliable climatic information (FRITTS, 1972). In this study a similar procedure has been adopted.

The aging effect is not yet clear but may be caused by intrinsic age-related physiological effects and/or local canopy effects, these effects are also observed for the $\delta^{13}C$ values (see LIPP et al., 1991; FRANCEY & FARQUHAR, 1982). The large δ^2H variations which sometimes occur from one year to another indicate that there may be other factors which affect the climatic signal. The discrepancy in the average δ^2H values of the individual trees may be caused by differences in genotype, the trees were not from the same clone.

4.1   Extraction of the climatic δ^2H signal and its comparison with meteorological data

To correlate the δ^2H values of late wood cellulose with meteorological data, it was necessary to evaluate the period for late wood synthesis. We found by dendrological means that the late wood in the "Schußbach" is mainly formed during July and August (unpublished results). In Table 1, linear regression analyses were carried out between δ^2H values of late wood cellulose and meteorological parameters (monthly temperature, relative humidity, and precipitation rate). It was found that temperature and humidity prevailing during July and August show the best correlation with isotope data of the contemporary year. The climatic parameters of other months are poorly correlated with δ^2H values (Table 1).

Thus, it is quite evident that the direct effect of temperature and relative humidity in the late wood growing season on the isotopic fractionation occurring during photosynthesis is a factor in determining the D/H ratios of late wood cellulose in tree-rings; the warmer and drier the summer climate, the higher the δ^2H values in late wood cellulose.

While there is evidence that the stable ^2H-isotopes in tree-rings carry a certain climatic signal, it is also evident from the poor correlation coefficients in Table 1 that this signal may be superimposed with non-climatic variations due to factors such as, for example, the "age trends" of the trees. The juvenile rings were omitted from these analyses (see above).

Table 1 Correlation of original average δ^2H values of late wood cellulose nitrate from 5 spruce trees with monthly temperatures, relative humidities and precipitation rates for the period from 1950 to 1981

Month	r_{temp}	r_{hum}	r_{prec}
Jan	0.27	0.31	0.03
Feb	0.24	0.21	-0.14
Mar	-0.06	0.10	-0.04
Apr	0.04	-0.09	-0.08
May	0.26	-0.22	-0.18
Jun	0.30	-0.20	-0.09
Jul	0.50 (S!)	-0.50 (S!)	-0.30
Aug	0.50 (S!)	-0.51 (S!)	-0.48 (S!)
Sep	0.07	-0.02	0.03
Oct	-0.07	-0.11	-0.28
Nov	0.05	0.25	0.30
Dec	-0.30	0.24	-0.20

S! = significant at 95% level (t-test)

Correlation of original, average late wood cellulose δ^2H values with July/August temperatures during 1950-1981 result in a relation $\delta^2H=128.5+4.3T$ with a correlation coefficient $r=0.61$. Correcting for the aging trends by correcting the δ^2H values of the mature rings (see Fig. 1) to obtain an average slope of zero, the δ^2H-temperature relationship is hardly improved (correlation coefficient $r = 0.63$; $\delta^2H = -133.7 + 4.4T$). Moreover, the relationships of δ^2H-relative humidity and δ^2H-precipitation rate are not improved using the correction technique (see Table 2).

Another source of scatter in the δ^2H-climatic parameter relationship may arise because in summers with a normal temperature and water regime, the δ^2H signal of late wood does not strongly respond to climatic variations. However, in summers showing distinct weather anomalies, which occur at the site mostly as "warm and dry" or as "cool and moist" ("signature years"), δ^2H values of the tree-ring late wood show a common increasing or decreasing trend ("100% signatures", see Fig. 1). Considering the "100% signatures" leads to improvement of the δ^2H-climatic parameter relationships as shown in Table 2. From these relationships it is obvious that a warm and dry (or cool and moist) summer climate will cause increased (or decreased) δ^2H-values in the late wood cellulose of tree-rings.

The temperature-δ^2H relationship is shown in Fig. 2, which presents correlations between local summer (July/August) temperatures and D/H ratios of both July/August precipitation

Table 2 Relationships of original, age-corrected, and signature δ^2H late wood average values of 5 trees (see text) with temperature, relative humidity, and precipitation rate in July/August.

	Original	Corrected	100% Signatures
Temp.	$\delta^2 H = -128.5+4.3T$ $r = 0.61$ (S!)	$\delta^2 H = -133.7+4.4T$ $r = 0.63$ (S!)	$\delta^2 H = -162.0+6.0T$ $r = 0.76$ (S!)
Rel. Hum.	$\delta^2 H = 10.7-0.9RH$ $r = -0.57$ (S!)	$\delta^2 H = 8.1-0.9RH$ $r = -0.58$ (S!)	$\delta^2 H = 25.2-1.2RH$ $r = -0.68$ (S!)
Prec. Rt.	$\delta^2 H = -49.7-0.1PR$ $r = -0.47$ (S!)	$\delta^2 H = -45.8-0.1PR$ $r = -0.50$ (S!)	$\delta^2 H = -41.5-0.25PR$ $r = -0.62$ (S!)

S! = significant on 95% level (t-test)

(solid line) and average D/H ratios of late wood cellulose (dashed line). An overall depletion of -20.7‰ in the δ^2H values of cellulose (dashed line) compared to the δ^2H values of precipitation (solid line) is observed. This may be caused by enzymatic fractionation during wood formation and agrees with values of EPSTEIN et al. (1976) who found an average fractionation of -22‰ between cellulose and associated meteoric waters. The slopes of the solid (precipitation) and the dashed line (cellulose) are very similar. This may signify a common climatic signal between temperature and both δ^2H of cellulose and δ^2H of precipitation. At a summer dry site, such as in "Schußbach", the July/August-precipitation may be used directly for formation of late wood cellulose. Thus, D/H ratios of late wood cellulose appear to reflect the summer (July/August) temperatures, which are in turn reflected in the D/H ratios of precipitation. The dashed-dotted line shows the relationship between temperature and average δ^2H values of the late wood cellulose from the "signature years" (see Fig. 1).

At the mentioned site it was observed that during extremely warm and dry years (e.g. 1976) the crowns of the trees dry up and only the lower parts survive. During such years drought has a limiting effect on plant growth, which is reflected in a significant increase in δ^2H values ("positive signature"). During years having propitiate climatic conditions (cool and moist) there are no limiting effects, which is reflected in a significant decrease in δ^2H-values ("negative signature"). Considering the "signatures" leads to improved expressions for the relationships between δ^2H-values and climatic conditions. The discrepancy of the solid (precipitation) and the dashed-dotted line (signatures) may be caused by the fact that even at shallow depth water can have a mean age of several months and consists of a mixture of actual precipitation and stored soil water. The D/H ratio of the soil water may undergo changes by means of evaporation, depending on summer temperature and humidity. In "Schußbach" there are often long periods of summer draught, which may cause δ^2H enrichment of the upper soil water and influence the δ^2H-value of cellulose. It is obvious that a complex system consisting of many variables is responsible for the δ^2H value of cellulose.

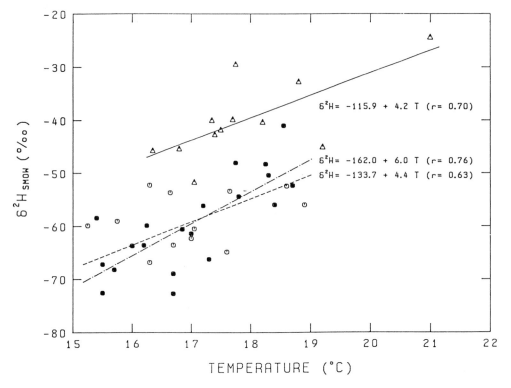

Fig. 2 Correlation of July/August temperatures measured at the meteorological station Neustadt/Aisch (315 m a.s.l.) with corrected (dashed line) and 100% signature δ^2H values (dashed-dotted line) of late wood cellulose during 1950-1981. The filled symbols refer to 100% signature years. The solid line shows the correlation between July/August temperatures and δ^2H values of summer (July/August) precipitation measured at the meteorological station Würzburg (182 m a.s.l.) 1978-1989

4.2 Comparison of δ^2H values of late wood cellulose and annual tree-ring width

Since the width of annual tree-rings have a certain climatic significance (BECKER & GLASER, 1991), they were measured and compared to δ^2H values of the tree-ring late wood cellulose (Fig. 3). As seen in Fig. 3, during most "signature years" a decrease of tree-ring

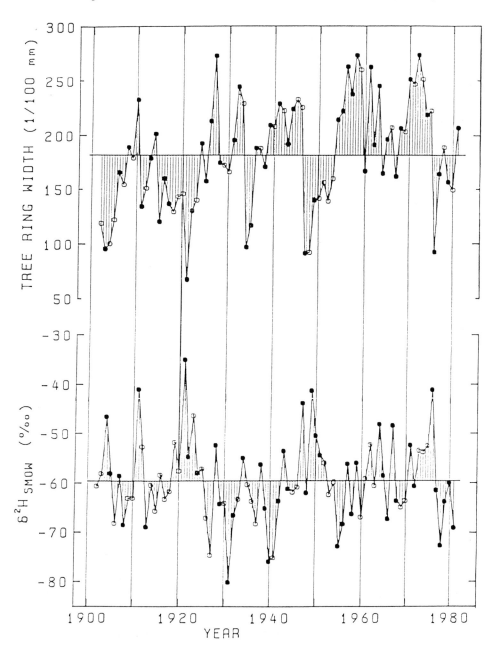

Fig. 3 Dendrometrical and isotopic dendrochronograms of spruce trees from the sampling area (Schußbach-Forest, FRG). Upper part: Annual tree-ring widths (average values) of five spruce trees. Lower part: δ^2H average values of late wood tree-ring cellulose from five spruce trees. The filled symbols refer to 100% signature years (see text)

width corresponds to an increase of the δ^2H value (negative correlation). During warm and dry "signature years" the limiting effect on plant growth is reflected in narrower tree-rings. In historical records the conspicuous years 1921, 1934, 1947, and 1976 are described as "hot and dry". The few exceptions (years 1932, 1949, 1956), which do not correspond with the above mentioned correlation, are probably caused by particular climatic conditions. This observation agrees with the increased number of tropical days (warm and moist) in 1956 [c.f. German Meteorological Service/Deutscher Wetterdienst] but cannot be interpreted further because of the lack of meteorological data. From the δ^2H values and tree-ring widths climatic conditions may be deduced. These methods provide a helpful tool in dating and interpreting climatic events in the present and past.

4.3   Rhythmic climate fluctuations during the twentieth century derived from δ^2H values of tree-ring cellulose

Figure 4 shows the data from Fig. 3 (lower part) after smoothing with a Gaussian low pass filter over 10 years. These values suggest the climatic variations indicated at the right vertical axis. Oscillations on different time scales are suggested by these data. Most dominant are oscillations with a periodicity of about 14 years, which are also evident from the auto-correlation function illustrated in the variance spectrum in Fig. 5. There is a significant periodicity at 14 years (see GRAF et al., 1990). Thus, the data indicate a dominant climatic oscillation occurring at about 14 years. We note that equivalent growth oscillations have also been observed by SCHWEINGRUBER et al., 1990. They found periods of abrupt tree-ring growth reductions in Central Europe with a frequency of 12 to 16 years, at an average of 14.5 years during the period 1850 to 1982. Also, in the Front Range Colorado, and in the Hudson Valley, New York intervals with an average of 13-14 years have been analysed. Normally, the growth reduction periods are not synchronized between the three areas investigated. However, these growth reductions investigated in spruce, fir, pine, and beech are essentially related to dry and warm periods during spring and summer months (SCHWEINGRUBER et al., 1990). The findings of SCHWEINGRUBER's and our research work support the conclusion that at least the local climate shows periodical oscillations which may be derived (semi-) quantitatively using isotopical tree-ring data.

5. Conclusions

The carbon-bound hydrogens of late wood cellulose from trees at a dry summer site obviously reflect the isotopic composition of the precipitation water used by the tree. In general, this in turn reflects local climate conditions in terms of the temperature prevailing during (summer) precipitation. Additionally, tree-ring δ^2H values reflect to some extent fluctuations of the relative humidity, and to a lesser extent fluctuations of precipitation rate. Our data strongly suggest that tree-ring widths and δ^2H values of late wood tree-ring cellulose

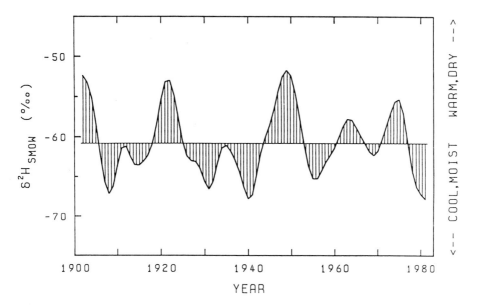

Fig. 4 Assignment of climatic parameters (temperature and humidity) to tree-ring δD-values during 1902-1981. The δD values are averaged over the five spruce trees and smoothed by a Gaussian low pass filter over 10 years. Dominant are climatic oscillations with a period of about 14 years

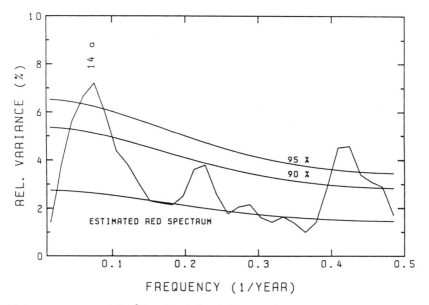

Fig. 5 Variance spectrum of the ^{2}H series deduced from the five spruce trees. The estimated red spectrum and confidence levels are calculated according to standard methods (SCHÖNWIESE, 1985). Oscillations with a period of about 14 years are significant at a 95% confidence level

reflect climate changes at a growth site with weather anomalies characterized either as "warm and dry" or as "cool and moist". During the twentieth century climatic oscillations in Central Europe appear to occur in 14-year cycles.

Of course, other climatic and non-climatic variables (e.g. cloudiness, radiation, wind, and soil water) may have certain (perhaps different) impacts on both, δ^2H-values and ring widths. This may explain the fact that δ^2H-values and ring widths sometimes do not show the same "signature" years (Fig. 3). Other variables influencing δ^2H values of cellulose and tree-ring widths should be further evaluated in the future.

References

ALEXANDER, W. J. & MITCHELL, R. L. (1949): Rapid measurement of cellulose viscosity by the nitration method. Analytical Chemistry 21, 1497-1500

BECKER, B. & GLASER, R. (1991): Baumringsignaturen und Wetteranomalien (Eichenbestand Guttenberger Forst, Klimastation Würzburg). Forstwiss. Cbl. 110, 66-83

BURK, R. L. & STUIVER, M. (1981): Oxygen isotope ratios in trees reflect mean annual temperature and humidity. Science 211, 1417-1419

DADSWELL, H. E. & HILLIS, W. E. (1962): Wood. In: Hillis,W.E. (ed.): Wood Extractives Vol. 3. Academic Press, New York, N.Y.

DANSGAARD, W. (1964): Stable isotopes in precipitation. Tellus 16, 436-468

EDWARDS, T. W. D. & FRITZ, P. (1986): Assessing meteoric water composition and relative humidity from ^{18}O and ^2H in wood cellulose: paleoclimatic implications for southern Ontario, Canada. Applied Geochemistry 1, 715-723

EPSTEIN, S.; THOMPSON, P. & YAPP, C. J. (1977): Oxygen and hydrogen isotopic ratios in plant cellulose. Science 198, 1209-1215

EPSTEIN, S.; YAPP, C. J. & HALL, J. H. (1976): The determination of the D/H ratio of non-exchangeable hydrogen in cellulose extracted from aquatic and land plants. Earth Planet. Sci. Lett. 30, 241-251

FRANCEY, R. J. & FARQUHAR, G. D. (1982): An explanation of $^{13}C/^{12}C$ variations in tree-rings. Nature 297, 28-31

FRITTS, H. C. (1972): Tree-rings and climate. Sci. Am. 227, 91-100

GRAF, W.; REINWARTH, O. & MOSER, H. (1990): The 520-year temperature record of a 100 m core from the Ronne ice shelf, Antarctica. Ann. Glaciol. 14, 90-93

GRAY, J. & SONG, S. J. (1984): Climatic implications of the natural variations of D/H ratios in tree-ring cellulose. Earth Planet. Sci. Lett. 70, 129-138

GRAY, J. & THOMPSON, P. (1976): Climatic information from $^{18}O/^{16}O$ ratios of cellulose in tree-rings. Nature 262, 481

KOZLOWSKI, R. (1962): Tree Growth. Ronald Press, New York, N.Y.

LIBBY, L. M. & PANDOLFI, L. J. (1974): Temperature dependence of isotope ratios in tree-rings. Proc. Nat. Acad. Sci. USA 71, 2482-2494

LIPP, J. & TRIMBORN, P. (1991): Long-term records and basic principles of tree-ring isotope data with emphasis on local environmental conditions. Paläoklimaforschung / Palaeoclimate Research 6, Gustav Fischer Verlag, Stuttgart, Jena, New York, 105-117

LIPP, J.; TRIMBORN, P.; FRITZ, P.; MOSER, H.; BECKER, B. & FRENZEL, B. (1991): Stable isotopes in tree-ring cellulose and climatic change. Tellus 43B, 322-330

NORTHFELT, D. W.; DE NIRO, M. J. & EPSTEIN, S. (1981): Hydrogen and carbon isotopic ratios of the cellulose nitrate and saponifiable lipid fraction prepared from annual growth rings of California redwood. Geochim. Cosmochim. Acta 45, 1895-1898

SCHIEGL, W. E. (1974): Climatic significance of deuterium abundance in growth rings of Picea. Nature 251, 582-584

SCHÖNWIESE, C. D. (1985): Praktische Statistik. Gebr. Bornträger, Berlin, 231 p.

SCHWEINGRUBER, F. H.; AELLEN-RUMO, K.; WEBER, U. & WEHRLI, U. (1990): Rhythmic growth fluctuations in forest trees of central Europe and the Front Range in Colorado. Trees 4, 99-106

STUMP, R. K. & FRAZER, J. W. (1973): Simultaneous determination of carbon, hydrogen and nitrogen in organic compounds. Nucl. Sci. Abstr. 28, 746

WARDROP, A. B. (1951): Cell wall organization and the properties of the xylem. Aust. J. Biol. Sci. 4B, 391

WILSON, A. T. & GRINSTED, M. J. (1977): $^{12}C/^{13}C$ in cellulose and lignin as palaeo-thermometers. Nature 265, 133-135

YAPP, C. J. & EPSTEIN, S. (1982a): Climatic significance of the hydrogen isotope ratios in tree cellulose. Nature 297, 636-639

YAPP, C. J. & EPSTEIN, S. (1982b): A reexamination of cellulose carbon-bound hydrogen δ^2H measurements and some factors affecting plant-water relationships. Geochem. Cosmochem. Acta 46, 955-965

YURTSEVER, Y. & GAT, J. R. (1981): Atmospheric waters. In: IAEA (ed.): Stable isotope hydrology. Deuterium and oxygen-18 in the water cycle, IAEA Technical Reports 210, 103-143

Authors' addresses:

Dr. J. Lipp, Institut für Hydrologie, GSF-Forschungszentrum für Umwelt und Gesundheit, Ingolstädter Landstr. 1, D-85764 Oberschleißheim
Dipl.-Phys. P. Trimborn, Institut für Hydrologie, GSF-Forschungszentrum für Umwelt und Gesundheit, Ingolstädter Landstr. 1, D-85764 Oberschleißheim

Tree-ring evidence of surface temperature variation during the past 1000 years

Malcolm K. Hughes

Summary

Six published reconstructions of seasonal temperatures, based on long tree-ring chronologies, are analysed for association with variation in solar activity inferred from fluctuation in the production of radiocarbon. No consistent association is found. Warmer conditions are reconstructed in the mid-twelfth century in several cases, and all show cold conditions in the early and mid-seventeenth century. A group of high-quality tree-ring based summer temperature reconstructions for the period A.D. 1740 to the late twentieth century was analysed. No consistent difference was found between the summer temperatures of three-year periods centred on the maxima and minima of the Zurich sunspot number series.

Résumé

Six reconstructions des températures saisonnières, effectuées à partir de longues séries dendrochronologiques et provenant de travaux publiés, ont été analysées afin de déterminer leur degré d'association avec les variations de l'activité solaire induites à partir des fluctuations de la production de radiocarbone. Aucune relation consistante ne fut observé. Dans plusieurs cas, les reconstructions climatiques montrent des conditions plus chaudes pour le milieu du XIIe siècle et dans tous les cas, elles indiquent que le début et le milieu du XVIIe siècle furent plus froids. Plusieurs reconstructions dendroclimatiques des températures estivales furent également analysées en relation avec le cycle des tâches solaires pour la période de 1740 A.D. à la fin du XXe siècle. Aucune différence consistante ne fut, toutefois, observé avec les températures estivales pour les trois années centrées sur les minima et maxima de la série des tâches solaires.

1. Introduction

1.1 General

A number of reconstructions of surface temperature based on tree-rings and covering the past 1000 years are now available. These may be of special significance in the study of the relationships between solar output and climate because they are series of annual estimates of the surface temperature in some season, usually summer. Not only are they of sufficient temporal resolution to detect fluctuations over a few years, such as might arise if variation between solar cycles affects surface climate, but they are continuous through the whole period. In the first part of this paper I will present some simple analyses of six published reconstructions 1000 years or more long, in order to demonstrate their main features relevant to the topic of this volume. Then, I will present analyses of a number of much shorter tree-ring based environmental records, covering between 250 and 400 years, and overlapping the observed record of solar activity.

1.2   Characteristics of tree-ring based temperature reconstructions

HUGHES & DIAZ (1994) have listed the characteristics of an optimum palaeoenvironmental record of climate over the last thousand years. These are: (1) continuity through the whole millenium and preferably longer; (2) dating good to the calendar year; (3) well understood spatial applicability; (4) a strong and well-defined climate signal of known seasonal applicability, and (5) temporal resolution of one year or better; (6) time-invariance in strength and nature of climate signals; and (7) the ability to record century-scale as well as interannual and decadal scale fluctuations. All six reconstructions discussed in Section 2 of this paper extend through the whole of the millenium, or longer, are precisely crossdated to the calendar year, and have a temporal resolution in their climate signal of a year or better.

Table 1   The six temperature reconstructions discussed in Section 2

Code	Region	Latitude	Longitude	Season	Standard deviation (deg.C)	Correlation /years/ stations[1]	Source
POL	Polar Urals	65N	70E	JJ	1.34	0.75/87/2	Graybill & Shiyatov, 1992
FEN	N.Fenno-scandia	65 and 70N	10,20,30E	AMJJA	0.64	0.79/111/4	Briffa et al, 1992b
SNS	Sierra Nevada,Ca.	36N	116W	JJA	0.97	0.44/119/21	Graumlich, 1993
SNW	Sierra Nevada	36N	116W	SONDJF	0.36	0.25/118/21	Graybill, 1993
TAS	Tasmania	42S	142E	NDJFA	0.40	0.44/114/4	Cook et al, 1992
RIA	Rio Alerce, Argentina	41S	71W	DJFMA	0.57	0.57/53/4	Villalba, 1990

The authors from whose works these reconstructions have been drawn (Table 1) address the strength, nature and seasonal applicability of their climate signals. In some cases, there is an explicit discussion of the spatial applicability of the climate signal (e.g., GRAYBILL, 1993), and they all give at least the locations of the trees sampled. Quantitative verification of the temperature reconstruction against independent or withheld instrumental data is carried out in all cases, using clearly described statistical techniques. Most of the authors also address the question of time-invariance of climate response. Similar comments apply to the shorter tree-ring based series dealt with in Section 3.

The question of whether low-frequency, i.e. century-scale, trends and fluctuations are accurately captured presents difficulties. With very few exceptions (e.g. HUGHES et al., 1984) the temporal stability of the statistical transfer functions is tested ("verified") against less, often much less, than 100 years of independent or withheld meteorological data. Thus, this verification does not help determine how well fluctuation on time scales longer than a few decades is recorded in the reconstructions. Departures from an accurate representation on the longer time scales might come from four principal sources. First, the number of cores (radial series) and trees making up each chronology varies through time. This is not a significant problem for the last 1000 years with the six long reconstructions discussed here. The number of chronologies used in the reconstruction may also vary through time. Some of the reconstructions discussed here are based on single site chronologies (TAS, RIA). Multiple-site mean chronologies for a small region are used for some (POL, FEN), and a third category of reconstruction is based on groups of separate site chronologies (SNS, SNW). HUGHES & DIAZ (1994) point out that reconstructions based on regional mean chronologies, or on groups of separate site chronologies, are much less likely to be distorted by non-climatic changes than those based on a single site chronology, since changes unique to a single site are to some degree diluted when material from multiple locations is used. Second, climate signal strength and nature may change in association with tree biology and ecology. This problem is dealt with more or less successfully according to the methods used to "standardize" the raw tree-ring measurements prior to their being combined in a site chronology (FRITTS, 1976; HUGHES et al., 1982; COOK & KAIRIUKSTIS, 1990). In most of the six reconstructions discussed here, considerable trouble has been taken to avoid removing information on low-frequency climate variation from the original series. In some cases the trees are the better part of a millenium old, or older (e.g. GRAUMLICH, 1993; GRAYBILL, 1993) and biological trend has been removed very conservatively. In others (e.g. BRIFFA et al., 1992b), an attempt has been made to identify the regionally characteristic age-related growth trend for the species involved, and only remove that. Third, site conditions for tree growth might change, due, for example, to gradual changes in soil chemistry or hydrology, or more rapid changes due to fire or wind throw. The influence of sudden disturbances can be minimized using standard dendrochronological techniques of sample selection. More gradual site changes are difficult to rule out, but use of material from several sites and species can help reduce their influence. Fourth, low-frequency shifts in the relative frequency of particular combinations of climate conditions affecting tree growth may distort the record. Selecting material from sites where tree growth is overwhelmingly dominated by a single limiting environmental factor has been the primary method used to avoid this problem. An alternate approach has been used by GRAUMLICH (1993) using response surfaces to represent non-linear interactions between climate factors. More extensive discussion of these issues may be found in FRITTS (1976), HUGHES et al. (1982) and COOK & KAIRIUKSTIS (1990).

2. Long reconstructions

Details of the six long reconstructions are given in Table 1. All are based on standardized tree-ring widths, with the exception of that for Fennoscandia (FEN) in which maximum latewood density is used. One is for a winter season temperature (SNW) but all others are for spring and summer months. All have been statistically verified, but in order to help the reader compare the strength of temperature signal, correlations with regional mean instrumental temperatures are given. Since the reconstructions are shown in all graphs as z-scores calculated for the common period A.D. 971 to 1970, the standard deviation of each reconstruction is also given in Table 1. The common period was defined by the first usable year of the shortest reconstruction and the last usable year of that ending first.

It is possible to see some persistent departures from the 1000-year mean in Fig. 1a. Examples of this include cold periods lasting twenty or more years in the twelfth and sixteenth centuries in FEN, and a warm period in the twelfth century in SNW and SNS. No consistent pattern of cold or warmth that may be associated with known or inferred periods of high or low solar output is evident. Much the same is true of Fig. 1b in which the reconstructions are shown as decadal means. More consistent pattern is evident, however, in the 30-year means (Fig. 1c), calculated because 30-year periods have been used as the basis for climatological comparisons in the descriptive literature. The early seventeenth century has cold 30-year periods in all reconstructions. The mid-twelfth century is warm in all reconstructions except POL, which has its warmest 30-year period of the millenium in the next period, covering the late twelfth, and early thirteenth centuries. Similar patterns can be seen in the century means (Fig. 1d). It should, however, be noted that the range of z-scores decreases from circa 4 to 1.5 to 0.9 to 0.4 in Figs. 1a to 1d, indicating that in most cases about 25% of variance is at wavelengths of 30 years or more, and only 10% or less at the century time scale, with the exception of FEN. The level of multidecade to century-scale variance in the other five reconstructions is consistent with the results of analysis of Manley's Central England instrumental temperature series for the period A.D. 1721-1990 (HUGHES & DIAZ, 1994).

The distribution of the 100 warmest and 100 coolest reconstructed seasons is shown in Fig. 2 for 30-year and 100-year periods. There appears to be a relative absence of unusually warm seasons and a relative abundance of unusually cold seasons in the seventeenth century in all six reconstructions. Some periods in some reconstructions show a relatively high abundance of both warm and cold years, for example the fourteenth century in POL, TAS and RIA, or conversely, a dearth of both, as in the late twelfth century in RIA, and the early twentieth century in SNS.

Annual

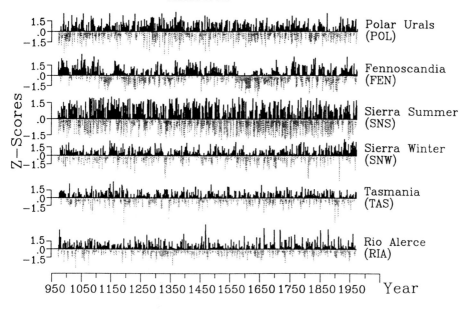

Decadal Means

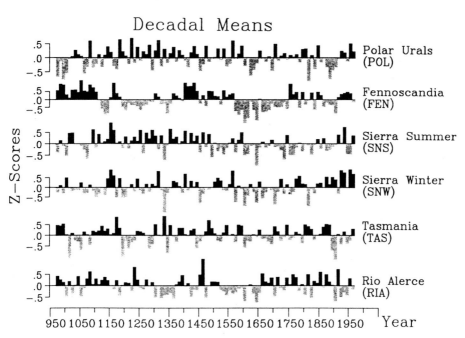

Fig. 1a The six temperature reconstructions listed in Table 1, shown as z-scores calculated for the period A.D. 971-1970: a) single years; b) decadal means

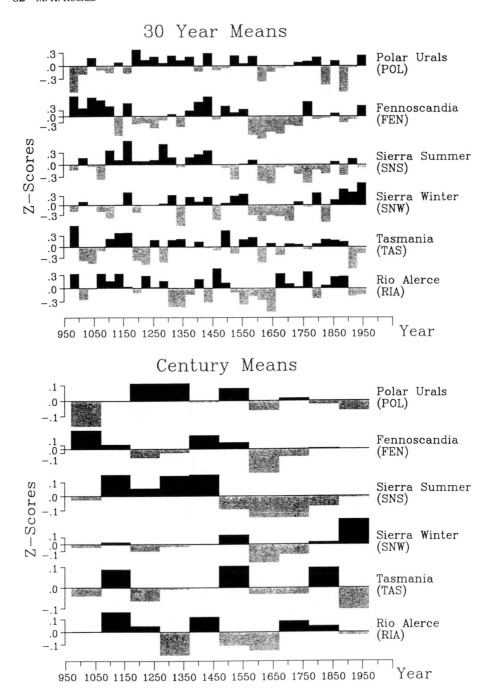

Fig. 1b   The six temperature reconstructions listed in Table 1, shown as z-scores calculated for the period A.D. 971-1970: c) 30-year means and d) hundred-year means

Distribution of Extreme Seasons

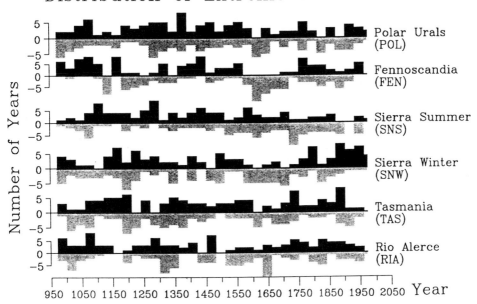

Distribution of Extreme Seasons

Fig. 2 Temporal distribution of the 100 warmest (black) and 100 coolest (grey) reconstructed seasons in each reconstruction listed in Table 1: a) 30-year periods; b) hundred-year periods

3. Shorter series

3.1 Solar maxima and minima

A group of well calibrated and verified reconstructions of surface temperatures from various regions of the northern hemisphere was selected (Table 2). Mean and standard deviations were calculated for reconstructed seasonal temperatures for the three years centred on each maximum or minimum in the Zurich sunspot numbers. The differences between the triplets centred on solar minima and those centred on solar maxima are slight and inconsistent. This is not changed by removing from consideration periods within three years of peaks in the Volcanic Explosivity Index (VEI) (BRADLEY & JONES, 1992), or containing years i and i+1 of severe El Niño events (QUINN & NEAL, 1992), or either.

Table 2 Means and standard deviations of years of various temperature reconstructions with high and low sunspot activity. Calculated from A.D.1740 to end of each series

Reconstruction/ months/source	Mean temperature of reconstructed seasons for three-year periods centered on sunspot maxima. (+/- 1 standard deviation)	Mean temperature of reconstructed seasons for three-year periods centered on sunspot minima. (+/- 1 standard deviation)
Kashmir AMJ/AS (Hughes and Shao, unpublished data)	19.04 +/- 0.81	19.13 +/- 0.36
Colorado Plateau [2] AMJJAS (Briffa et al, 1992a)	-0.085 +/- 0.40	-0.104 +/- 0.29
Pacific NW AMJJAS[3] (Briffa et al, 1992a)	-0.065 +/- 0.35	-0.084 +/- 0.28
Polar Urals JJ (Graybill and Shiyatov, 1992)	-0.579 +/- 1.31	-0.366 +/- 0.87
Edinburgh, U.K., JA (Hughes et al, 1984)	14.49 +/- 0.83	14.42 +/- 0.42
Sierra Nevada, Calif., JJA (Graumlich, 1993)	-0.399 +/- 1.024	-0.219 +/- 0.56

[2] The mean of grid-points 15 and 16 (see BRIFFA et al, 1992a)   [3] The mean of grid-points 17 and 18 (see BRIFFA et al, 1992a)

4. Discussion

If we were dealing with six meteorological stations that had been in operation for a thousand years, it would be unlikely that they could capture global temperature changes on multidecadal to century time scales, unless those changes were persistent and large enough to force a change in mean and dispersion of temperatures in all or most of the six regions in the same direction. It is clear from the instrumental record that changes in global mean surface temperature of a degree Celsius or so, persistent over more than one year (for example following the 1992 eruption of Mount Pinatubo), do not result in all regions' temperatures varying in the same direction. In fact, the records used here are more or less imperfect reconstructions of past temperatures, the best of which might portray 50% of the variance of the actual temperature variations, and whose ability to represent accurately fluctuations on multidecadal and century time scales is not well known. Notwithstanding all this, these are among the best records available for the past 1000 years, and they do show some interesting features. They do not provide support for a simple "Medieval Warm Period / Little Ice Age" sequence, although there are indications that the early and mid-seventeenth century was colder than average in all, with a concentration of unusually cold seasons and a relative lack of unusually warm seasons. This period is before the Maunder Minimum in solar activity recorded from the late-seventeenth, early eighteenth century. There is no apparent association between these temperature reconstructions and other features of solar activity variation inferred from the radiocarbon record, for example the Wolf and Spörer Minima. There is some evidence for warmer conditions at several of these sites in the mid-twelfth century, coinciding with the inferred peak of the Medieval Solar Maximum (JIRIKOVIC & DAMON, 1994).

Since the Zurich sunspot series is available for more than 250 years, it has been possible to compare mean summer temperatures of triplets of years centred on maxima and minima in this series for a number of well-calibrated temperature reconstructions. No consistent pattern of differences was found, even after years that might have been influenced by volcanic events, severe El Niño events, or either, had been eliminated.

Although the analyses presented have been very simple, the data analysed are some of the best available for the past 1000 years. The results of these analyses provide no support for a clear and direct relationship between multidecadal and century-scale fluctuations in solar receipts and mean seasonal surface temperatures. They do cast light on the nature of temperature fluctuations on these time scales, and lend some support for the existence of a widespread cold period in the mid-seventeenth century, and perhaps a warm period in the mid-twelfth century. More definitive analyses must await the development of many more such records, with at least as strong a signal as in these reconstructions. Clearly, the longer these new records are, the better the possibility of detecting any systematic century-scale fluctuations.

Acknowledgements

I am grateful to the Institute of Physics, Bologna University, for the opportunity to attend this ESF workshop. The work reported here was done during my tenure of a Visiting Fellowship at the Cooperative Institute for Research in Environmental Science at the University of Colorado, Boulder. I am grateful to the several authors who have kindly made their published reconstructions available in convenient electronic form. Peter Brown, Gregg Garfin and Jon Eischeid gave invaluable help with the analyses. Jacques Tardif kindly translated the summary. This work was supported in part by National Science Foundation

Grant ATM-90-11048. This is a contribution to project ARRCC "Analysis of recent and rapid climate change".

References

BRADLEY, R. S. & JONES, P. D. (1992): Records of explosive volcanic eruptions over the last 500 years. In: Bradley, R. S. & Jones, P. D. (eds.): Climate Since A.D. 1500. London, Routledge, 606-622

BRIFFA, K. R.; JONES, P. D. & SCHWEINGRUBER, F. H. (1988): Summer temperature patterns over Europe: a reconstruction from 1750 A.D. based on maximum latewood density indices of conifers. Quat. Res. 30, 36-52

BRIFFA, K. R.; JONES, P. D. & SCHWEINGRUBER, F. H. (1992a): Tree-ring density reconstruction of summer temperature patterns across western North America since 1600. J. Climate 5, 735-754

BRIFFA, K. R.; JONES, P. D.; BARTHOLIN, T. S.; ECKSTEIN, D.; SCHWEINGRUBER, F. H.; KARLÉN, W.; ZETTERBERG, P. & ERONEN, M. (1992b): Fennoscandian summers from A.D. 500: temperature changes on short and long time scales. Climate Dynamics 7, 111-119

COOK, E.; BIRD, T.; PETERSON, M.; BARBETTI, M.; BUCKLEY, B.; D'ARRIGO, R.; FRANCEY, R. & TANS, P. (1991): Climatic change in Tasmania inferred from a 1089-year tree-ring chronology of Huon pine. Science 253, 1266-1268

COOK, E.; BIRD, T.; PETERSON, M.; BARBETTI, M.; BUCKLEY, B.; D'ARRIGO, R. & FRANCEY, R. (1992): Climatic change over the last millennium in Tasmania reconstructed from tree-rings. The Holocene 2, 205-217

COOK, E. R.; BIRD, T.; PETERSON, M.; BARBETTI, M.; BUCKLEY, B. & FRANCEY, R. (1992): The Little Ice Age in Tasmanian tree-rings. In: Mikami, T. (ed.): Proc. Int. Symp. LIA. Tokyo Metropolitan Univ., 11-17

COOK, E. R. & KAIRIUKSTIS, L. A. (1990): Methods of dendrochronology: applications in the environmental sciences. Kluwer Academic Publishers, Boston, 394 p.

FRITTS, H. C. (1976): Tree-rings and climate. Academic Press, London, 567 p.

GRAUMLICH, L. J. (1993): A 1000-year record of temperature and precipitation in the Sierra Nevada. Quat. Res. 39, 249-255

GRAYBILL, D. A. (1993): Dendroclimatic reconstructions during the past millennium in the southern Sierra Nevada and Owens Valley, California. In: Lavenburg, R. (ed.): Southern California Climate: Trends and extremes of the past 2000 years. Natural History Museum of Los Angeles County, Los Angeles, CA

GRAYBILL, D. A. & SHIYATOV, S. G. (1992): Dendroclimatic evidence from the northern Soviet Union. In: Bradley, R. S. & Jones, P. D. (eds.): Climate Since A.D. 1500. London, Routledge, 393-414

HUGHES, M. K.; KELLY, P. M.; PILCHER, J. R. & LAMARCHE, V. C. Jr. (eds.) (1982): Climate from tree-rings. Cambridge Univ. Press, Cambridge, 223 p.

HUGHES, M. K.; SCHWEINGRUBER, F. H.; CARTWRIGHT, D. & KELLY, P. M. (1984): July-
August temperature at Edinburgh between 1721 and 1975 from tree-ring density and
width data. Nature 308, 341-344

HUGHES, M. K. & DIAZ, H. F. (1994): Was there a Medieval Warm Period, and, if so,
where and when? Climatic Change 26, 109-142

JIRIKOVIC, J. L. & DAMON, P. E. (1994): The Medieval solar activity maximum. Climatic
Change 26, 309-316

QUINN, W. H. & NEAL, V. T. (1992): The historical record of El Niño events. In: Bradley,
R. S. & Jones, P. D. (eds.): Climate Since A.D. 1500. London, Routledge, 623-648

VILLALBA, R. (1990): Climatic fluctuations in northern Patagonia in the last 1000 years as
inferred from tree-ring records. Quat. Res. 34, 346-360

Author's address:

Prof. Dr. M. K. Hughes, Laboratory of Tree-ring Research, University of Arizona, USA-Tucson, AZ
85721 (address for mail), and Cooperative Institute for Research in Environmental Science,
University of Colorado, USA-Boulder, CO 80309

Century-scale solar oscillation in terrestrial reservoirs

Giuliana Cini Castagnoli, Giuseppe Bonino & Antonello Provenzale

Summary

A new fundamental oscillatory mode with frequency υ_0 of about 2.4×10^{-3} yr^{-1} has been discovered by STUIVER & BRAZIUNAS (1989) in the analysis of the 9600-year long, high-precision record of the atmospheric ^{14}C production rate. This radiocarbon mode was suggested to be related to changes which possibly occur in the sun's convective zone. The main evidence for this radiocarbon component in the ^{14}C spectrum is the presence of energetic peaks at frequency υ_0 and higher harmonics. In this paper we show that similar spectral features are detected in the 1600-year long thermoluminescence (TL) profile of the GT14 sea sediment core. Two energetic spectral peaks, significant at the 95% level, are present in the TL spectrum at $3\upsilon_0$ and $7\upsilon_0$. The frequencies $2\upsilon_0$ and $4\upsilon_0$ appear respectively as the frequency of the modulation amplitude and the inverse of the node separation of a modulated wave train generated by the beat of two decennial components, both significant at the 95% level, which dominate the high-frequency part of the TL spectrum. These results indicate that different terrestrial reservoirs apparently resonate on the same fundamental frequencies and suggest that they may be driven by partially solar and/or similar climatic mechanisms.

Zusammenfassung

Bei der Analyse der einen Zeitraum von 9600 Jahren umfassenden Präzisionsaufzeichnungen der atmosphärischen ^{14}C-Produktion entdeckten STUIVER & BRAZIUNAS (1989) eine neue Basisschwingung mit der Frequenz υ_0 von etwa 2.4×10^{-3} pro Jahr. Diese Radiokarbonproduktionsrate wird mit Änderungen in Verbindung gebracht, die möglicherweise in der Konvektionszone der Sonne auftreten. Der wichtigste Hinweis auf die Radiokarbonkomponente im ^{14}C-Spektrum ist das Auftreten von Spitzen im Energiespektrum bei der Frequenz υ_0 und in Vielfachen davon. In dem vorliegenden Beitrag wird dargelegt, daß ähnliche spektrale Eigenschaften in den 1600-Jahres-Thermoluminiszenz-profilen des Meeresbodensedimentkerns GT14 nachgewiesen werden können. Zwei energetische spektrale Spitzen - signifikant auf dem 95%-Niveau - sind im TL-Spektrum bei $3\upsilon_0$ und $7\upsilon_0$ vorhanden. Die Frequenzen $2\upsilon_0$ und $4\upsilon_0$ erscheinen jeweils als Frequenz der Modulationsamplitude bzw. der Inverse der nodalen Trennung eines modulierten Wellenzuges, die durch zwei Komponenten im Zehnjahresbereich erzeugt werden; sie sind signifikant auf dem 95%-Niveau erkennbar, das den Hochfrequenzteil des TL-Spektrums dominiert. Diese Ergebnisse zeigen, daß die verschiedenen terrestrischen Reservoire offenbar auf denselben Hauptfrequenzen mitschwingen und so die Vermutung erlauben, daß sie teilweise durch solare und/oder ähnliche klimatische Mechanismen angetrieben werden.

1. Introduction

The thermoluminescence (TL) measured in different depths of the GT14 sea sediment core comes from its polymineralic crystals (mainly quartz and calcite). Generally speaking, the grains which constitute the mud of the core should have an initial TL level which reflects the equilibrium between the environmental ionizing radiations and direct light effects, acting in competition. In recent sediments the post-depositional TL acquired *in situ*, due to irradiation from the local radioactivity concentration in the core, plays a minor role and does not mask the pre-depositional TL signal. Since the sediment was never exposed to light after extraction, the analysis of the TL profile of the GT14 core provides information on the equilibrium TL level at the time of deposition of the grains. Careful analysis of the spectral properties of the TL profile of the core has been described elsewhere (Cini Castagnoli et al., 1988a,b, 1989, 1990, 1991). In the above papers we have also shown that common spectral features are detected in the TL profile, in the sunspot number series and in other terrestrial records of solar activity (such as the radiocarbon record). Here we discuss only the comparison between the TL profile, obtained by readings on the glow curves at 340°C, and the radiocarbon signal found by Stuiver & Braziunas (1989).

We recall the main characteristics of the analysis of the GT14 core. This sediment is a 1.17 m long carbonate mud core which was drilled in a water depth of 166 m on the continental shelf in the Gulf of Taranto near Gallipoli (Italy), at 39°45'55"N, 17°53'30"E. The core has been sampled at equal intervals Δd=2.5 mm for a total of 467 layers. Four TL glow curves per layer were taken on 15 mg samples of sieved material (composed of particles with size < 44 μm). A fifth glow curve per layer was measured after bleaching one sample with a sun-lamp for 420 minutes. Well-defined, regular oscillations are detected in the natural TL profiles, while no periodic oscillations are present in the bleached signal.

The use of the ^{210}Pb method and a careful tephroanalysis of known volcanic events, together with the information deduced from the presence of a ^{137}Cs peak at the top of the sediment, allowed for a precise determination of the sedimentation rate (Bonino et al., 1993). The sample interval Δd has been found to correspond to a time interval Δt=3.87 years and the TL depth profiles have thus been transformed into TL time series. Further details on these issues may be found in Cini Castagnoli et al. (1988a,b, 1990, 1991).

2. Results and discussion

Figure 1 reports a detrended TL profile of the GT14 core (in the inset) and the corresponding power spectral density. The TL signal shown in the inset has been obtained by averaging the four natural TL profiles (readings on the glow curves at 340°C) and by removing a 160-year running average in order to avoid the effects of long-term trends (CINI CASTAG-NOLI et al., 1988b). Since we have four replicas of the TL record, the 95% significance level in the spectrum is given by $1/2\,P(\upsilon) \leq \pi(\upsilon) \leq 4P(\upsilon)$, where $\pi(\upsilon)$ is the "true" spectrum and $P(\upsilon)$ is our spectral estimate. Four significant spectral peaks are evident in the TL spectrum, with periods 137.7 yr, 59 yr, 12.06 yr and 10.8 yr.

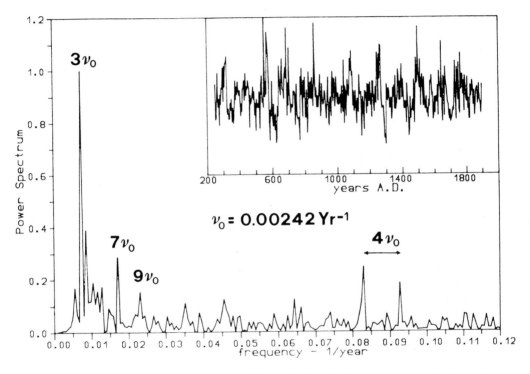

Fig. 1 Power spectral density of the detrended TL signal. The TL data are shown in the inset. The frequencies of the main spectral peaks are indicated above the spectrum

Recalling that the fundamental frequency in the radiocarbon series, as indicated by STUIVER & BRAZIUNAS (1989), is $\upsilon_0 \sim 2.4 \times 10^{-3}$ yr^{-1}, one realizes that this frequency also plays an important role in the TL data. The most energetic peaks at 137.7 yr and 59 yr are in fact at the frequencies $3\upsilon_0$ and $7\upsilon_0$. Table 1 reports the periods of the main odd harmonics of υ_0 recorded in the ^{14}C data and in the TL series. Similar periodicities in the ^{14}C series in tree-rings and in the carbonate profiles measured in Ionian Sea cores were found by CINI CASTAGNOLI et al. (1992).

The even harmonics of υ_0 with frequency $2\upsilon_0$ and $4\upsilon_0$ are also important in the TL series, since they are respectively found as the frequency of the amplitude modulation and the inverse of the node distance of a modulated wave train which is generated by the beat of the two high-frequency components of 12.06 yr and 10.8 yr. The amplitude modulation of the wave train has a period of T=206 years (i.e. frequency $\sim 2\upsilon_0$), while the carrier has a period of 11.4 years, a value close to the periodicity of the eleven-year sunspot cycle. We finally recall that the absence of a 206-year peak in the TL spectrum is not due to the effects of having removed the long-term trend from the data, since no peak with approximately this frequency is present in the spectrum of the raw (non-detrended) data. We also note that, because of the length of the record, the frequency resolution of the TL spectrum in the long-term range (~ 400 yr) is too poor to univoquely determine the presence or the absence of a peak at the fundamental frequency υ_0 in the TL signal.

STUIVER & BRAZIUNAS (1989) suggested that the oscillation with frequency $\upsilon_0 \sim 2.4 \times 10^{-3}$ yr^{-1}, discovered in the analysis of the ^{14}C record, may be a fundamental secular mode of the sun. In this paper we have shown that the same fundamental frequency υ_0 also accounts for the main spectral features of the TL profile of the GT14 sediment core. These results seem to indicate that different terrestrial reservoirs resonate at the same frequencies and suggest common solar and/or climatic origins for the two records considered.

Table 1 Periods of the main odd harmonics of the fundamental frequency υ_0 in the ^{14}C record and the TL profile

	^{14}C record	TL profile
υ_0	2.4×10^{-3} yr^{-1}	2.42×10^{-3} yr^{-1}
3rd harm	T=138.9 yr	T=137.7 yr
5th harm	T=83.3 yr	T=82.6 yr
7th harm	T=59.5 yr	T=59.0 yr
9th harm	T=46.3 yr	T=43.6 yr

References

BONINO, G.; CINI CASTAGNOLI, G.; CALLEGARI, E. & ZHOU, G. M. (1993): Radiometric and tephroanalysis dating of recent Ionian Sea cores. Nuovo Cimento 16C, 155-162

CINI CASTAGNOLI, G.; BONINO, G. & PROVENZALE, A. (1988a): On the thermoluminescene profile of an Ionian Sea sediment: evidence of 137, 118, 12.1 and 10.8-year cycles in the last two millenia. Nuovo Cimento 11C, 1-12

CINI CASTAGNOLI, G.; BONINO, G. & PROVENZALE, A. (1988b): The thermoluminescence profile of a recent sedimentary core and the solar variability. Sol. Phys. 117, 187-197

CINI CASTAGNOLI, G.; BONINO, G. & PROVENZALE, A. (1989): The 206-year cycle in tree-ring radiocarbon data and in the thermoluminescence profile of a recent sea sediment. J. Geophys. Res. 99, 11971-11976

CINI CASTAGNOLI, G.; BONINO, G.; PROVENZALE, A. & SERIO, M. (1990): On the solar origin of the thermoluminescence profile of the GT14 core. Sol. Phys. 127, 357-377

CINI CASTAGNOLI, G.; BONINO, G. & PROVENZALE, A. (1991): Solar-terrestrial relationships in recent sea sediments. In: Sonnett, C. P. & Gianpapa, M. S. (eds.): The Sun in Time. Univ. of Arizona Press, 562-586

CINI CASTAGNOLI, G.; BONINO, G.; SERIO, M. & SONETT, C. P. (1992): Common spectral features in the 5500-year record of total carbonate in sea sediments and radiocarbon in tree-rings. Radiocarbon 34, 798-805

STUIVER, M. & BRAZIUNAS, T. F. (1989): Atmospheric ^{14}C and century-scale solar oscillations. Nature 338, 405-408

Author's addresses:

Prof. Dr. G. Cini Castagnoli, Istituto di Cosmogeofisica del CNR, Corso Fiume 4, I-10133 Torino, and Istituto di Fisica Generale dell'Università, Via P. Giuvia 1, I-10125 Torino

Prof. Dr. G. Bonino, Istituto di Cosmogeofisica del CNR, Corso Fiume 4, I-10133 Torino, and Istituto di Fisica Generale dell'Università, Via P. Giuvia 1, I-10125 Torino

Dr. A. Provenzale, Istituto di Cosmogeofisica del CNR, Corso Fiume 4, I-10133 Torino

Medieval mud and the Maunder Minimum

Claudio Vita-Finzi

Summary

In the search for solar-climatic relations the alluvial record can usefully complement palaeoclimatic data because it bears on changes in seasonality and latitudinal gradients. Channel aggradation between A.D. 300 and A.D. 1850 in the Old World between 45° and 25° N and between 10° W and 60° E reflects more equable discharges. Its time-transgressive onset and close point to progressive shifts in the cyclonic belt. The Maunder Minimum of solar activity (A.D. 1645-1715), sometimes blamed for the Little Ice Age of Europe, does not fit the chronology. Moreover an increase, rather than a decrease, in solar energy, could lead to changes in UV absorption in the lower stratosphere, displacements in the polar vortex and jet stream, and the replacement of meridional by zonal regimes. The latest long-term maximum in solar activity to be manifested in the tree-ring ^{14}C record peaked in about A.D. 600.

Résumé

La chronologie des alluvions peut fournir à l'étude des rapports soleil-climat des indices utiles et complémentaires aux données paléoclimatiques car elle revèle soit les changements saisonnières soit les variations progressives qui correspondent à la latitude. L'alluvionnement qui a eu lieu entre 300 et 1850 après J.-C. dans l'ancien monde entre 45° et 25° Nord et 10° Ouest et 60° Est indique des apports fluviatiles moins irréguliers qu'aujourd'hui; le diachronisme de son début et de son achèvement indique le déplacement progressif vers le Nord et vers le Sud de la zone cyclonale. Le minimum des tâches solaires Maunder (1645-1715) est quelquefois associé avec le Petit Age Glaciaire en Europe mais il ne correspond pas du tout bien à sa chronologie. D'ailleurs un accroissement plutôt qu'une diminution de l'energie solaire pourrait modifier l'absorption du rayonnement solaire ultraviolet dans la basse atmosphère, des déplacements du tourbillon polaire et du courant-jet, et la substitution des régimes méridionaux par les régimes zonaux. Le dernier maximum à grande échelle de l'activité solaire témoigne par le contenu en ^{14}C des anneaux de croissance des arbres a fait une pointe environ 600 après J.-C.

1. Introduction

The influence of changes in solar irradiance on global climate remains elusive: thus, whereas REID (1991) found a close correspondence between global sea surface temperature and total irradiance as derived from the envelope of the sunspot cycle since 1860, WIGLEY & RAPER (1990) were able to show that it had had little effect on global-mean temperature since 1874. The Little Ice Age and earlier cool periods of the Holocene (GROVE, 1988) are also not explicable by irradiance changes (WIGLEY & KELLY, 1990).

A major obstacle to advancing the analysis is the brevity of dependable records of solar activity. It can be circumvented by using the cosmogenic isotopes ^{10}Be and ^{14}C as inverse measures of the solar wind. The approach has attracted most attention in the study of solar cycles, and agreement is reported between short-term radiocarbon variations and the width of Bristlecone tree-rings (SONETT & SUESS, 1984) and the fluctuations of Mono Lake in California (STINE, 1990).

The problem is made more tractable still by focusing on a distributional rather than a numerical effect. During solar cycle 21 (1976-1986) the greatest variations occurred at ultraviolet and radio wavelengths (LEAN, 1991). A rise in UV radiation promotes ozone reduction and hence stratospheric warming, as well as a displacement of the polar vortex and the jet stream (LANDSCHEIDT, 1987). The issue now becomes one of identifying shifts in the general circulation rather than in temperature (or some other variable) at individual locations.

2. Solar activity

Much of the research into solar influence on climate has considered the ~11 year sunspot cycle and its derivatives. REID (1991), for example, used the 11-year running mean Zurich sunspot number for his analysis. FRIIS-CHRISTENSEN & LASSEN (1991) relied on variations in the length of the solar cycle, short cycles being associated with high solar activity. And, thanks mainly to the work of EDDY (1976, 1988), the Spörer (A.D. 1420-1530), Maunder (A.D. 1645-1715) and other sunspot minima are seen as manifestations of even more sluggish oscillations.

The use of radiocarbon records as a surrogate measure of solar activity and hence as a way of extending the historical sunspot sequence into pre-observational times is complicated by the possible shielding effect of the geomagnetic field and by climatic changes especially through their effects on the oceanic circulation (SIEGENTHALER et al., 1980). A plot of atmospheric Δ^{14}C values based on tree-ring data for the last 9700 years shows a 14% decrease from about 6000 B.C. to A.D. 500 followed by a rise of some 2% (SONETT & FINNEY, 1990). The smoothed line approximates a 400-year moving average (STUIVER et al., 1991). Atmospheric Δ^{14}C values calculated from dipole intensities derived from remanent magnetism data after setting equilibrium Δ^{14}C values prior to 11,000 yr B.P. at 90 per mil give a good match with the measured results (STUIVER et al., 1991). On the other hand there is poor agreement between the predicted effects of geomagnetic modulation and the [10]Be record (OESCHGER & BEER, 1990). Parallelism between the two isotopic narratives, though complicated by different mixing and depositional mechanisms, would point to cosmic (and thus solar) control.

In short, long-term solar modulation remains "an unproven but viable option" (STUIVER et al., 1991) for explaining the ^{14}C trough: the "astonishing" increase in ^{14}C and [10]Be after A.D. 500 can be ascribed to a decrease in the earth's magnetic field; but other possibilities include meteorological effects on [10]Be deposition, unknown carbon cycle effects, and "an unusual summation of periods of a quiet sun" (OESCHGER & BEER, 1990).

3. Alluvial chronologies

Alluvial sequences, including the periods of non-deposition they embody, are datable manifestations of changes in the hydrological behaviour and sediment production of river basins. They thus provide a summary account of a catchment, though evidently on a much more modest scale than isotopic sequences from ocean or ice sheet cores, a feature that permits changes in the seasonal pattern of discharge to be recovered from the sedimentary record provided there is a source deposit to serve as standard of comparison for the sediment under review.

Between about A.D. 500 and A.D. 1850 a phase of channel aggradation affected many valleys in an area extending at least from Morocco to the Makran and the Sahara to the Dordogne (Fig. 1). Its chronology depends on archaeological, historical and calibrated radiocarbon ages. The values listed in Table 1 (Fig. 2) were determined on charcoal. Although charcoal is subject to fluvial transport its fragility makes it less prone to long-delayed redeposition than most other sources of ^{14}C ages and when present in the substantial amounts required for conventional ^{14}C dating it commonly is *in situ*. Archaeological and historical limiting ages, as well as the location of the charcoal samples within the sections, are discussed in the sources cited.

The channel fill is characterized by well stratified, often drab, clay-poor alluvium over a basal gravel often derived from an older, poorly sorted alluvial and colluvial channel fill rich in reddish iron oxides; it contains mineralogical and organic evidence of waterlogged conditions (MACLEOD & VITA-FINZI, 1982), and has led to a steepening of channel long profiles since reversed by channel incision. Conflicting dates for periods of slope erosion or for interludes of delta growth do not invalidate the observation that there was widespread aggradation about A.D. 300-1850 in many valleys previously and subsequently affected by channel incision.

Aggradation, though complicated by land use and tectonics, was primarily due to a change in stream discharges leading to silt-clay depletion, the prevalence of reducing conditions,

Table 1 ^{14}C dates on charcoal from historical valley fill

LOCATION	LAT	AGE	(YR BP)	LAB NO.[b]	SOURCE
		UNCOR-RECTED	CALIB[a]		
ITALY					
Fiano Romano	42°12'	1350±100	1290±110	I-4802	1
Fiano Romano	42°12'	1670±95	1585±165	I-4801	1
Valchetta	42°00'	1140±160	1025±245	I-881	1
Crescenza	42°00'	1400±100	1305±85	I-882	1
Moncalieri	44°59'	790±50	695±45	R-618A	14
Moncalieri	44°59'	900±50	815±105	R-618	14
C. Monferrato	45°10'	1595±50	1520±105	R-622	14
C. Monferrato	45°10'	1580±50	1430±105	R-622A	14
Treia	41°52'	1675±45	1585±105	I-6109	2
Treia	41°52'	490±30	525±10	I-6110	2
FRANCE					
Manaurie	44°57'	1335±90	1285±110	I-6794	3
MOROCCO					
W. Mellah	33°37'	490±90	525±95	I-2693	1
W. Bou Regreg	33°58'	800±200	705±240	Ł-398A	1
W. Beth	34°30'	280±60	310±120	GrN-2198	1
SPAIN					
Ayna	38°31'	820±35	730±40	SSR-726	4
Ayna	38°31'	850±50	755±120	SSR-727	4
Ayna	38°36'	780±110	690±100	SSR-728	4
Ayna	38°29'	420±65	505±50	SSR-729	4

Table 1 continued

IRAN					
Khorramabad	33°30'	330±105	360±150	I-3214	1
H. Langi	27°21'	110±90	< 285	HAR-1114	5
H. Langi	27°21'	460±170	515±205	HAR-1707	5
H. Langi	27°21'	930±80	830±120	HAR-1706	5
QATAR					
Ras Abaruk	25°43'	600±90	585±85	HAR 639	6
Bir Markhiyah	25°43'	1850±90	1820±185	HAR 638	6
GREECE					
Voidomatis	40°00'	1000±50	940±140	OxA-191	7
Voidamatis	40°00'	800+100	705±115	OxA-192	7
PORTUGAL					
Boi	37°50'	780±50	690±45	BETA 2909	8
Arade	37°18'	2690±70	2780±135	OxA-2267	13
CYPRUS					
Vasilikos	34°50'	470±80	520±35	n/a	9
TUNISIA					
W. Akarit	37°58'	1470±190	1365±195	n/a	10
W. Akarit	37°58'	610±110	585±85	n/a	11
LIBYA					
W. Ganima	32°42'	610±110	585±85	Q-656	1
ALGERIA					
El Abadia	36°17'	700±80	670±105	MC-2530	12
El Abadia	36°17'	120±70	< 280	BETA-2678	12
El Abadia	36°17'	270±70	305±305	MC-2531	12

equable discharges and a steepening of channel gradients. In view of the hydrological variability of the area affected, the most economical explanation is reduced rainfall seasonality abetted to varying degrees by human activity. In short, palaeohydrology can be employed as a surrogate for atmospheric circulation.

The original version of Fig. 2 displayed a latitudinal gradient for both the onset and the close of aggradation, which prompted the suggestion that deposition had accompanied a progressive southward shift in depression tracks which was supplanted by incision as the depressions shifted back northwards (VITA-FINZI, 1976). As with the major climatic episodes of the Quaternary the onset of aggradation is more gradual than its close.

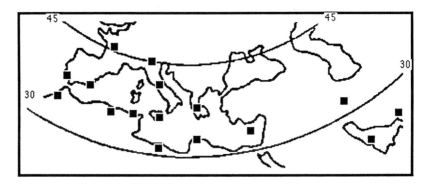

Fig. 1   Location of sections listed in Table 1

The new graph omits the limiting archaeological ages that were an important feature of the first draft in favour of new radiocarbon ages determined by other workers in the intervening years. Nevertheless the 1976 age envelope is retained (Fig. 2) as a test of the latitudinal thesis. The new ages fall comfortably within it with two exceptions: that obtained for a basin (rather than a channel) fill at Bir Markhiyah, Qatar, and that for a deposit in Portugal which CHESTER & JAMES (1991) in fact correlate with a unit (Odiaxere Group) deposited after ca. 2000 yr B.P.

4. The climatic link

The need remains for a mechanism to account for a general shift in hydrological regimes over a large, climatically inhomogeneous region. LAMB (1964) had inferred a southward displacement of the main depression tracks by 5-10° in the European sector to explain the climatic changes of A.D. 1200-1300 to 1600-1700. Such a displacement would moderate the seasonal nature of discharge regimes in the drainage basins of the Mediterranean and the Near East by replacing a predominantly zonal by a meridional circulation pattern (BUCHA, 1983).

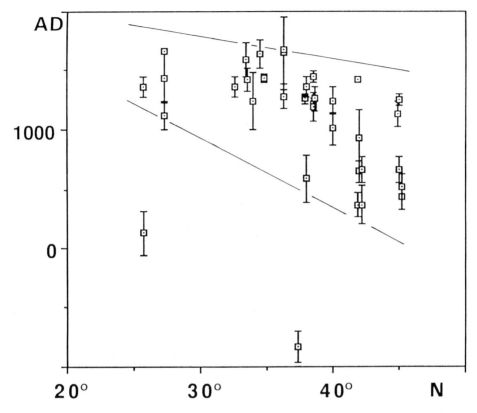

Fig. 2   Age/latitude plot of ^{14}C-dated valley fills. Data from Table 1. The envelope indicates the zone encompassed by the latitudinal lag of VITA-FINZI (1976) proposed when only fourteen radiocarbon ages were available. The outliers are discussed in the text

The records show that the prevailing cyclonic tracks in the North Atlantic are displaced southward at periods of sunspot maximum by up to 3° in the North Sea area (HERMAN & GOLDBERG, 1978). Maxima in solar activity are marked by increased UV and microwave radiation (LEAN, 1991), whose effect on the mesosphere and upper stratosphere is to promote ozone reduction and hence stratospheric warming, and by a displacement of the polar vortex and of the jet stream (LANDSCHEIDT, 1987).

The proposed mechanism runs counter to any causal association between the Little Ice Age and the Maunder Minimum. Quite apart from the chronological misfit - the effect preceded the supposed cause by over a century - the change in solar activity is in the wrong direction. The hydrological excursion that laid down the Medieval mud was a response to the cumulative rise in solar irradiance that peaked in about A.D. 600 and that was briefly interrupted by the Oort, Wolf, Spörer, Maunder and Dalton spells of sunspot minima.

The observation by Bucha (1983) that meridional circulation patterns are associated with times of low geomagnetic activity might suggest that the ^{14}C record cannot discriminate between increased transmission and reduced reception of solar energy. But rapid advances in the study of palaeomagnetism should soon reduce the ambiguity and make possible further comparisons between solar and terrestrial history.

Acknowledgments

I thank Emily Lincoln for help and Antonio Veggiani for valuable discussions.

References

Bucha, V. (1983): Direct relations between solar activity and atmospheric circulation, its effect on changes of weather and climate. Stud. Geophys. Geod. 27, 19-45

Chester, D. K. & James, P. A. (1991): Holocene alluviation in the Algarve, southern Portugal: the case for an anthropogenic cause. J. Archaeol. Sci. 18, 73-87

Devereux, C. M. (1983): Recent erosion and sedimentation in southern Portugal. Ph.D. thesis, London Univ., 283 p.

Eddy, J. A. (1976): The Maunder Minimum. Science 192, 1189-1202

Eddy, J. A. (1988): Variability of the present and ancient sun: a test of solar uniformitarianism. In: Stephenson, F.R. & Wolfendale, A.W. (eds.): Secular solar and geomagnetic variations in the last 10,000 years. NATO ASI Series 236, Kluwer, Dordrecht, 1-23

Fontes, J.-C. & Gasse, F. (1989): On the age of humid Holocene in late Pleistocene phases in North Africa. Palaeogeogr. Palaeoclimatol. Palaeoecol. 70, 393-398

Friis-Christensen, E. & Lassen, K. (1991): Length of the solar cycle, an indicator of solar activity closely associated with climate. Science 254, 698-700

Gomez, B. (1987): The alluvial terraces and fills of the Lower Vasilikos Valley, in the vicinity of Kalavasos, Cyprus. Trans. Inst. Brit. Geogr. 12, 345-359

Grove, J. M. (1988): The Little Ice Age. Methuen, London

Herman, J. R. & Goldberg, R. A. (1978): Sun, Weather and Climate. NASA, Washington (reprinted Dover, New York 1985)

Lamb, H. H. (1964): Climatic changes and variations in the atmospheric and ocean circulations. Geol. Rdsch. 54, 486-504

Landscheidt, T. (1987): Long-range forecasts of solar cycles and climatic change. In: Rampino, M. R.; Sanders, J. E. & Königsson, L. K. (eds.): Climate history, periodicity, and predictability. Van Nostrand Reinhold, New York, 421-445

Lean, J. (1991): Variations in the sun's radiative output. Rev. Geophys. 29, 505-535

MacLeod, D. A. & Vita-Finzi C. (1982): Environment and provenance in the development of recent alluvial deposits in Epirus, NW Greece. Earth Surf. Proc. Landf. 7, 29-43

Oeschger, H. & Beer, J. (1990): The past 5000 years history of solar modulation of cosmic radiation from ^{10}Be and ^{14}C studies. Phil. Trans. Roy. Soc. London A 330, 471-480

PAGE, W. D. (1972): The geological setting of the archaeological site at Oued el Akarit and the palaeoclimatic significance of gypsum soils, southern Tunisia. Ph.D. thesis, Univ. of Colorado, 111 p.

REID, G. C. (1991): Solar total irradiance variations and the global sea surface temperature record. J. Geophys. Res. 96, 2835-2844

SIEGENTHALER, U.; HEIMANN, M. & OESCHGER, H. (1980): ^{14}C variations caused by changes in the global carbon cycle. Radiocarbon 22, 177-191

SONETT, C. P. & FINNEY, S. A. (1990): The spectrum of radiocarbon. Phil. Trans. Roy. Soc. London A 330, 413-426

SONETT, C. P. & SUESS, H. E. (1984): Correlation of Bristlecone pine ring widths with atmospheric ^{14}C variations, a climate-Sun relation. Nature 307, 141-143

STINE, S. (1990): Late Holocene fluctuations of Mono Lake, eastern California. Palaeogeogr. Palaeoclimatol. Palaeoecol. 78, 333-381

STUIVER, M. & BECKER, B. (1986): High-precision decadal calibration of the radiocarbon time scale, A.D. 1950-2500 B.C. Radiocarbon 28, 863-910

STUIVER, M.; BRAZIUNAS, T. F.; BECKER, B. & KROMER, B. (1991): Climatic, solar, oceanic and geomagnetic influences on Lateglacial and Holocene atmospheric ^{14}C/^{12}C change. Quat. Res. 35, 1-24

THOMMERET, Y.; KING, G. C. P. & VITA-FINZI, C. (1983): Chronology and development of the 1980 earthquake at El Asnam (Algeria), a postscript. Earth Planet. Sci. Lett. 63, 137-138

VITA-FINZI, C. (1974): Age of valley deposits in Périgord. Nature 250, 568-570

VITA-FINZI, C. (1975a): Chronology and implications of Holocene alluvial history of the Mediterranean Basin. Biul. Geol. 19, 137-147

VITA-FINZI, C. (1975b): Late Quaternary deposits of Italy. In: Squyres, C. (ed.): Geology of Italy. Tripoli, Libya, 329-340

VITA-FINZI, C. (1976): Diachronism in Old World alluvial sequences. Nature 263, 218-219

VITA-FINZI, C. (1978): Environmental history. In: de Cardi, B. (ed.): Quat. Archaeological Report. Oxford Univ. Press, Oxford, 11-25

VITA-FINZI, C. & GHORASHI, M. (1978): A recent faulting episode in the Iranian Makran. Tectonophysics 44, T21-T25

WIGLEY, T. M. L. & KELLY, P. M. (1990): Holocene climatic change. Phil. Trans. Roy. Soc. London A 330, 547-560

WIGLEY, T. M. L. & RAPER, S. C. B. (1990): Climatic change due to solar irradiance changes. Geophys. Res. Lett. 17, 2169-2172

WOODWARD, J. C.; LEWIN, J. & MACKLIN, M. G. (1992): Alluvial sediment sources in a glaciated catchment, the Voidomatis Basin, northwest Greece. Earth Surf. Proc. Landf. 17, 205-216

Author's address:

Prof. Dr. C. Vita-Finzi, University College London, Department of Geological Sciences, Gower Street, GB-London WC1E 6BT

Climatic features during the Spörer and Maunder Minima

Dario Camuffo & Silvia Enzi

Summary

The first aim of this article is to increase knowledge about climate by looking for and analysing written sources over a long period of time. If the earth system's response to the climatic forcing agents is known exactly in the various climatic epochs, on the basis of documentary evidence taken from the past, then reliable models can be constructed for forecasting future scenarios. The second objective of this paper is to determine whether there is any relationship between climatic changes and variation in solar activity, in order to ascertain whether there really is a cause-effect relationship as suggested by EDDY (1977). From the analysis of some long series of significant climatic events (flooding of the Po and some of its tributaries, flooding of the Tiber and the Adige, the sea surges and freezing over of the Venice lagoon, strong wind, storms, hailstorms, heavy rains, plagues of locusts), some basic conclusions have been drawn, among which are the following: The period between 500-1000 A.D., held to be the worst in terms of climate, should be treated cautiously in that there are many gaps in the documentation; the same with reference to the tenth and eleventh centuries. From the cooling down in the Little Ice Age, it can be seen that changes in the levels of the oceans was not an immediate, simple and unequivocal response to changes in temperature. The earth-sun relationship can only be interpreted if the climatic periods which occurred during the minima of sunspot activity, are inserted into the complete climatic context. EDDY's hypothesis should be proven on the basis of the earth's behaviour in every period of solar irregularity. From the data gathered and analysed, it would seem that the theory is not reliable, in that the climate during the Spörer Minimum was extremely unstable, while in the Wolf and Maunder Minima it was not. It would appear that during a period of climatic abnormalities, the Spörer Minimum occurred casually enough, at the same time. This observation is further supported by the fact that, in some cases, the climatic abnormalities preceded the solar ones. The really exceptional phenomena, such as the great winters, can occur at any time, even during epochs of global warming.

Zusammenfassung

Das vorrangige Ziel dieses Artikels besteht darin, den vorhandenen Kenntnisstand über das Klima zu erweitern, indem über lange Zeiträume hinweg aufgezeichnete, schriftliche Quellen systematisch ausgewertet werden. Wenn es gelingt, mit Hilfe von historischen

Quellen die Reaktion der Erde auf unterschiedliche klimatische Steuerungsgrößen für verschiedene Zeit- bzw. Klimaabschnitte exakter zu erfassen, dann können auch zuverlässige Modelle für zukünftige Klimaszenarien erarbeitet werden. Ein weiteres Anliegen des vorliegenden Beitrages ist die Diskussion der Frage, ob sich klar faßbare Beziehungen zwischen Klimaveränderungen und Veränderungen in der Sonnenaktivität nachweisen lassen und damit zu überprüfen, ob hier tatsächlich eine Kausalbeziehung zwischen Ursache und Wirkung vorliegt, wie sie von EDDY (1977) postuliert wird. Basierend auf der Auswertung mehrerer langfristiger Aufzeichnungen über herausragende klimatische Ereignisse (Hochwässer des Tiber, Adige, Po und einiger seiner Nebenflüsse, marine Hochwässer vor Venedig, Zufrieren der Lagune von Venedig, Starkwinde, Stürme, Hagelstürme, Starkregen, Heuschreckenplagen) haben wir mehrere grundlegende Schlußfolgerungen gezogen, von denen die folgenden hier vorgestellt werden: Die Periode zwischen 500 und 1000 n.Chr., bislang für die klimatisch ungünstigste gehalten, sollte mit Vorsicht betrachtet werden, da es hier Lücken in der schriftlichen Dokumentation gibt; dasselbe gilt für das zehnte und elfte Jahrhundert. Am Beispiel der sogenannten "Kleinen Eiszeit" kann gezeigt werden, daß die zeitgleichen Meeresspiegelstände und -schwankungen keine unmittelbare, einfache und einheitliche Reaktion auf die Temperaturveränderungen darstellten. Die Beziehung Erde-Sonne kann nur interpretiert werden, wenn die Perioden, die sich durch ein Minimum der Sonnenflecken auszeichnen, in den gesamten klimatischen Kontext gestellt werden. So sollte EDDYs Hypothese auf der Grundlage der Reaktion der Erde auf jede Periode einer irregulären Sonnenaktivität überprüft werden. Im Hinblick auf die zusammengetragenen und analysierten Daten scheint EDDYs Theorie jedoch nicht haltbar zu sein, da das Klima beispielsweise während des Spörer Minimums extrem instabil war, im Gegensatz zu den Klimaverhältnissen während des Wolf und des Maunder Minimums. Es scheint, daß sich das Spörer Minimum während einer Periode klimatischer Abnormitäten eher zufällig etablierte. Diese Beobachtung wird durch die Tatsache untermauert, daß in einigen Fällen klimatische Anomalien den solaren vorangingen. Wirklich außergewöhnliche klimatische Phänomene wie die "Great Winters" können jederzeit auftreten, sogar während Perioden einer globalen Erwärmung.

1. Introduction

The climate, because of its very nature, is subject to continuous variations, some of which are slow while others are rapid. Many scenarios are known to have occurred in the past and models of future forecasts are based on reasonable hypotheses. Unfortunately, the complex mechanisms involved in the climatic processes, the earth system's time of response to the external and internal forcing as well as the natural and anthropogenic factors have not been integrated into these models. It is therefore important to determine which of the possible scenarios proposed for the future are the most reliable. The theoretical problem then becomes a practical one, in fact, it becomes a vital one to resolve for cities like Venice, which lies right on the sea coast that is exposed to various types of risks triggered off by atmospheric forcing agents, including the submersion of the whole town in the event of the expected temperature rise due to the greenhouse effect.

The first aim of this article is to increase knowledge about the climate, by recovering, screening and analysing the available information based on written sources, in order to understand what had really happened over a vast time scale. There had already been transitions from cold to hot periods and *vice versa*, and it was also possible to observe, from the same sources, the sea's reaction during those periods.

Over the last 2000 years, in fact, different climatic epochs have alternated, with the respective transitions periods. The early Middle Ages Warm Epoch (culminated 1000-1200), the Little Ice Age (1430-1850) and the Present-Day Warming (1850-today) are the most well known. Knowing what has happened in the past would help in understanding the earth system's response to the climatic forcing agents, thus being better able to determine what could happen in the future on the basis of documentary evidence, rather than on simple hypothetical grounds. The second objective of this paper is to investigate if there is any link between climatic changes and variations in solar activity, in order to verify whether there really is a cause-effect relationship. In fact, EDDY (1977) suggested that climatic deterioration in past centuries corresponded to periods of minimum solar activity, and therefore, derived from a reduced solar input. Solar influence would only emerge if the periods affected had really been anomalous with respect to the whole climatic context, and if such irregularities occurred each time solar activity had been anomalous. Having long series of environmental data available allows for the overall observation of the phenomenon, instead of having to study it in separate pieces, during some of the more critical periods. Thus, it is possible to distinguish between mere coincidence and a general rule. Fortunately, the time interval under consideration covers several periods in which the solar activity was minimum, that is the early Medieval Minimum (EMM), i.e. 660-740 A.D.; the Oort Minimum (OM), 1010-1090; the Wolf Minimum (WM), 1282-1342; the Spörer Minimum (SM), 1416-1534 and the Maunder Minimum (MM), 1645-1715. Some of the more representative phenomena will now be examined.

2. The climate during the early centuries of the Christian era

Greek and Latin literature furnish some very interesting indications (CAMUFFO, 1990), such as, for example, those given about the notorious year of 44 B.C. The period from 45 A.D. to 50 A.D. was characterized by drought, while the winters of the years 57/58, 68/69, 73/74 and 173/174 A.D. were particularly cold (but did not reach the terrible levels of those of the years 176/177, 274/275 and 398/399 B.C. when the Tiber froze). In the year 164 A.D. there were heavy snowfalls that later caused many floods. In the first century A.D., northern winds were rather rare in Rome and the marine breeze began about a month earlier than now. The first centuries A.D. were relatively calm, up until the year 589, when bad weather affected the whole of northern Italy, especially the Po Valley. The so-called "Paolo

Diacone flood" occurred in 589, named after the most famous of the chroniclers who described it. The flooding of the Po and its tributaries, as well as the Tiber, will be described in greater detail, from the early Christian era, in the following paragraph. All the series have been processed with the Tukey-Hanning filter (WEI, 1990), with period $T = 50$ years and steps $S = 5$ years.

3. Flooding of the Po and its tributaries

The first flooding of the Po is mentioned in 204 B.C. As in every long series, there is a wealth of information around the tenth and eleventh centuries. The apparent greater frequency of such events in the second millenium does not necessarily indicate a net worsening of the climate with respect to the past but should rather be explained in terms of greater available information. The series should then, logically, be divided into two parts: one covering reports from before the year 1000 and a second from 1000 on, when not only the more dramatic events were recorded, but also the minor ones.

In the first part, from 200 B.C. to the end of the first millenium (see Fig. 1a), there was a calm period from the second century A.D. almost until the fifth. From the elevated cultural conditions that dominated until the fall of the Roman Empire and the growing importance of Milan and, more generally, the whole of the Po Valley (the Emperor Maximianus chose Milan as his official residence in 286, Constantine also residing there), it should be surmised that during this period there were no hydrological catastrophes, rather it being a period devoid of information. There was a flood immediately before the EMM, but that period did not appear to be any worse than the preceding ones.

During the second millenium, the flooding began immediately after the OM, but this does not seem to be a hydrologically anomalous period. It seems, more simply, to be the beginning of a period when information was more precise and detailed, after the dark gap of the tenth and eleventh centuries. Two peaks dominate this period: one at the beginning of the 1300s and one straddling the 1500s, when the sunspot activity of the SM was at its lowest. During the MM however, there was a secondary, irrelevant maximum with respect to the fluctuation levels.

The auto-regressive maximum entropy (MEMSA) harmonic analysis (Fig. 1b) give a highly fluctuating graph without any dominant peaks. The highest shows a recurrence interval of little more than 400 years, with various lesser peaks at intervals of 30 years, the first peak also corresponds to this value.

4. Flooding of the Tiber

The various documentary sources about the Tiber (literary, historical, epigraphical, archives, etc.), cover an uninterrupted period of 24 centuries. The available series is substantially complete, but not homogeneous. The problems concerning the lack of homogeneity are common to all long series and derive from the changes affecting the territory, the banks and urban development in Rome. When, in fact, the city was not densely populated, as in the beginning, the houses were perched only on the highest part of the hills, so that floods reaching the houses corresponded to a much higher flood level than the flood level reaching the houses built on the lower slopes or in the valleys. The constriction of the bed, which occurred when parts of the river were taken to form parks in the villas built in the 1500s, altered the hydrological regime of the Tiber and the course of its flow through Rome. A similar problem, but due to the accumulation of debris, occurred in the Middle Ages, when the river was neglected, when the remains of uprooted trees and other debris lodged in the foundations of the crumbling bridges. For many centuries, waste was commonly discarded directly into the river, and it was also perfectly normal to throw all kinds of rubbish into it. Up until the last century, the mills situated along the river formed another obstacle to its course. The banks, too, changed with time, but essentially, they were only improved after the 1870 flood. Then, of course, there are the single histories of the tributaries, territory and the anthropogenic activity which have divided the series into smaller homogeneous subseries, covering each one for one or more centuries. Moreover, the series is, overall, extremely interesting, in that it shows how the different hydrological periods alternated over such long periods.

The first details about the flooding of the Tiber go back to 414 B.C., during the so-called Greek Minimum of solar activity (around 400 B.C.), but this was not a particularly singular period hydrologically, as far as the Tiber was concerned, even though there were droughts, very hard winters and violent storms. The worst period with respect to floods (Fig. 2a) occurred at the beginning of the second century B.C.; between the first century B.C. and the second century A.D. with a greater frequency at the extremes; especially in the first half of the first century B.C.; then in the interval between 500 and 700 A.D. There were then some more or less sporadic floods, with a gap between the tenth and eleventh centuries. Later, two really marked peaks occurred: one at the end of the 1400s (at the beginning of the SM) and another half way through the 1600s, precisely at about the time the MM was about to start. During the WM, the situation remained within normal limits, while the minimum solar activity coincided with a minimum of the fluctuations in the flood distribution. During the OM at the beginning of the millenium there do not appear to have been any floods, while the part concerning the second millenium starts immediately after it, although this may be irrelevant, in that it coincides with a period that was not very well documented. The EMM happened during a positive fluctuation in the series, but not in a period that can be considered unusual. Thus, it must be concluded that even in the case of the Tiber, the climatic anomalies were not due to the different changes in solar activity. In particular, during the SM and MM, the hydrological phenomenon preceded the astrophysical event, inverting the sequence that would be expected for a cause-effect relationship.

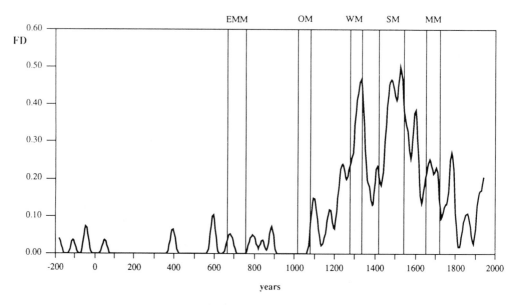

Fig. 1a Frequency distribution of the floods of the river Po and its tributaries (Tukey-Hanning window, T=50 years, S=5 years)

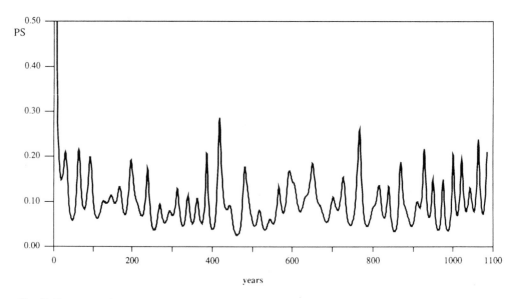

Fig. 1b Recurrence interval (yr) of the river Po and its tributaries (MEMSA)

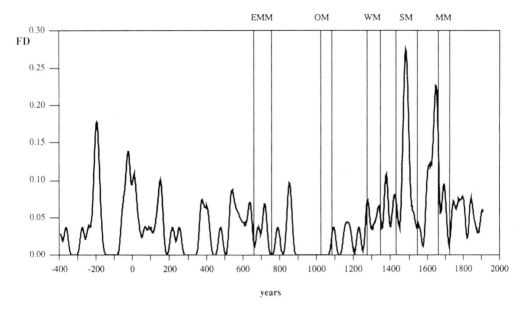

Fig. 2a Frequency distribution of the floods of the river Tiber (Tukey-Hanning window, T=50 years, S=5 years)

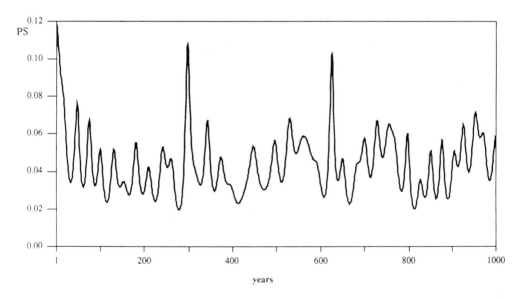

Fig. 2b Recurrence interval (yr) of the river Tiber (MEMSA)

The MEMSA spectral analysis (Fig. 2b) shows a very marked recurrence peak of 100 years (which follow the relative multiples); further peaks occur ever 30 years, even though the first 30-year peak is not evident, having been reabsorbed by the initial accumulation.

5. Flooding of the Adige

The floods of the river Adige (Fig. 3) have always followed a rather tortuous trend, but they became particularly frequent at the beginning of the 1500s, in the middle of the SM. They then occurred at intervals of approximately 120 years, which have already been commented upon. The second principal peak, in fact, occurred at the beginning of the MM and there was a third peak during the second half of the 1700s and a fourth at the end of the 1800s. It should be noted that during the SM there were, in effect, all sorts of natural calamities, while the MM was only occasionally troubled, passing, for the most part, without any natural disasters.

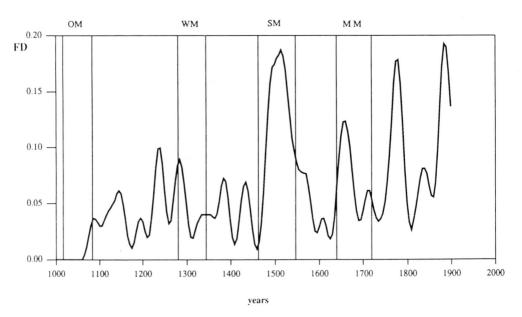

Fig. 3 Frequency distribution of the floods of the river Adige (Tukey-Hanning window, T=50 years, S=5 years)

Thus, if EDDY's hypothesis were correct, it seems strange that the earth-sun relationship was very strong between the 1400s and 1500s but weak between the 1600s and 1700s, leading to the conclusion that there is little, if any cause-effect relationship resulting from solar activity.

6. Sea storms at Venice

Sea storms at Venice (locally called "high water") are closely connected to the problems of sea level. The phenomenon is set off by the confluence of certain factors:

- the presence of a depression over the central-western Mediterranean, generating a Scirocco wind (SE) along the Adriatic basin and a Bora wind (NE) in the Upper Adriatic, which transport masses of water towards Venice;
- the overturned barometric effect that leads to the redistribution of the level in the Mediterranean, so that level rises where the atmospheric pressure falls;
- the formation of free water oscillations (seiches) in the Adriatic basin, with a resonance period approximating the tide's, following impulses originating from rapid changes in pressure.

The main astronomic forcing results from the combination of the lunar-solar forces, the periods being mainly linked to the earth's rotation, the lunar synodic month and lunar declination (18.6 years). The earth was in perihelion in A.D. 1250, but the maximum combination of the attracting forces (moon in perigee and earth in perihelion) occurred in 1424, in 1433 and in 1442. The next similar combination will occur in about 1300 years time, while the first minimum occurred some 1450 years ago (LAMB, 1972). This means that the lunar-solar forcing agent, apart from having a period of 18.6 years, was not the root cause of any anomalous period, except the one between the second quarter of the 1400s. Ground subsidence has also led to a progressive rise in the sea level, being particularly evident in the last century due to utilisation of the water-bed and to work carried out in the area (PIRAZZOLI, 1982; CARBOGNIN & TARONI, 1984; RUSCONI, 1992). Later in this paper, the most significant results of a recent study will be discussed (CAMUFFO, 1993).

Unlike the relatively brief period of astronomic anomaly, the high water phenomenon was therefore characterized by two main factors: atmospheric circulation (the arrival of Scirocco-type depressions) and the average sea level (eustacy). The former determined the level of fluctuation on a daily scale, the second, the average value of the sea, the former overlapping the latter, and, depending on the average level at the start, modest fluctuations may have been transformed into high water.

Comparing the seasonal distribution of high water during the period between 1200 and 1800 and in this century, it can be seen that it has hardly varied at all. It seems rather strange that the seasonal frequency of the Scirocco wind has not changed during the different climatic periods, in that the Scirocco transports air masses having high temperatures and humidity levels, and is accompanied by heavy rain and is, therefore, among the principal possible responsibles for climatic changes.

The 18.6-year period of lunar declination is well documented, as can be seen when using different filters: The MEMSA spectral analysis of the data (Fig. 4a) has shown that the most frequent recurrent intervals are 110, 300, 400 and 450 years, with a secondary peak every 25 years. It is interesting to note that the 110-year peak is very close to the 125-year peak found from the Padua long series on precipitation (1725-today; CAMUFFO, 1984) and the Roman one (1787-today; COLACINO & PURINI, 1986), or in the ^{14}C spectrum (CINI CASTAGNOLI et al., 1992).

The periods when the frequency of high water increased anomalously (Fig. 4b), are as follows:
-   1914 up to the present, when the frequency increased out of all proportion due to the combined effects of hydraulic work carried out in the lagoon (especially in the excavation of the canals) and at the mouth of the port, and because of subsidence. It is the most alarming situation that has occurred in the last millenium.
-   1720-1830, after the MM, which would demonstrate that there is no connection with the solar constant.
-   1500-1550, in the SM, which occurred at the initial phase of the Little Ice Age, and which some authors attribute to an anomaly in the solar constant.
-   1424-1442, a period of maximum lunar-solar attraction, which did have a certain influence, but it was not as great as the meteorological one.
-   1250-1350, (which is the WM), was part of a stormy period, starting as the earth moved out of the perihelion and which characterized the transition period after the end of the Middle Ages Warm Epoch. At the same time there were also violent storms in the North Sea and in the English Channel, which completely altered the coastline.

From the many documents on the time and cumulative distribution of high water frequency it can be concluded that during the long cooling-down period in the Little Ice Age, the sea level in Venice never dropped. However, this should have happened, according to the models which indicate that the sea level will rise greatly, due to the forecasted warming over the next few years, as a result of the greenhouse effect. What probably separates the models from reality is the time required for the system to respond. In fact, the earth system is composed of various heterogeneous parts where, for example, the response time by the Antarctic would be some thousands of years. The phase displacement with respect to the climatic forcing agents would be enormous, in terms of the action exerted by this continent both storing and releasing water. Thus, the actual response would be complex and difficult to determine and if it were based only on the immediate trends of forcing agents, it would not always be easy to forecast nor might it be accurate.

7. Freezing of the Lagoon

In the past, the extreme harshness of some winters has caused the lagoon to freeze over completely. The freezing of the edges, where the water level is low and there is little exchange with sea water, occurred fairly frequently. It was also fairly common to see sheets of ice, transported by the rivers, deposited directly into the lagoon. When the cold was more intense and lasting however, the lagoon was gripped in a vice of ice preventing the boats, with their cargoes of supplies, from reaching the shore. The problem became dramatic until the ice became thick enough to support the passage of people and carts etc., and supplies could reach the city from the mainland.

The lagoon freezes over when there is an intense period of cold that lasts over a sufficiently long time (at least 2-3 weeks). Snow in the area facilitates the phenomenon, in that the surface temperature of the "ground" is never above 0°C, not even in presence of solar radiation which is reflected, and the masses of cold air that flow over the surface of the lagoon are always cold. The cold Bora wind removes, moreover, considerably sensible and latent heat from the lagoon waters.

The hydraulic works carried out in the lagoon, above all the diversion of the canals into the sea and their excavation, the surrounding dams and the mouth of the port have altered the quantity of lagoon water that is exchanged with the sea, with the result that it would be more difficult, today, for the lagoon to freeze over. Notwithstanding this, the lagoon did freeze over, even recently, and in a climatic period that has, on average, been the warmest over the last 500 years.

An analysis of the documentary data (Figs. 5a,b; CAMUFFO, 1987, 1990) led to the following conclusions:
-   The maximum frequency of freezings occurred right at the end of the Little Ice Age, between 1700 and 1850.
-   There were many freezings in the fifteenth and sixteenth centuries, with a slight attenuation at the beginning of the 1500s.
-   The severity of the winters was considerably attenuated for almost a century in the 1600s during the Little Ice Age, this attenuation being more or less global, traces being also found in other parts of Europe and Japan.
-   With reference to a possible relationship with solar activity (EDDY, 1977), it should be noted that the absolute maximum frequency of freezings occurred in a period when there were no solar anomalies. The MM was a period of minimum frequency of freezings, while during the SM, many were documented, but not more so than in the successive 500 years. During the WM the lagoon never froze. This would lead to the conclusion that there is no correlation between these freezings and solar activity.
-   The distribution of "severe" winters (CAMUFFO & ENZI, 1992), shows a greater frequency in the Little Ice Age as would be logical to expect. In contrast, the really "great" winters (those winters when the maximum rigidity reached the maximum levels, as evaluated from the dramatic effects) were, for the most part, randomly distributed. One, in particular, occurred in 1929 during the terminal phase of the rapid global warming, and a second was verged on in 1985, when it was the duration rather than the intensity

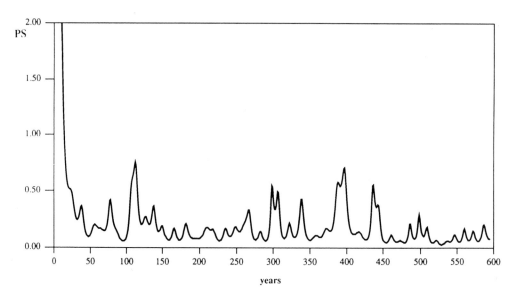

Fig. 4a Recurrence interval (yr) of the sea surges in Venice (MEMSA)

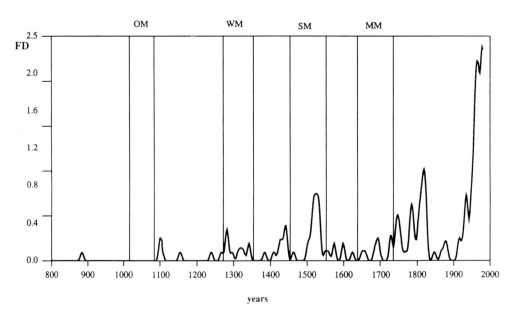

Fig. 4b Frequency distribution of the sea surges in Venice (Tukey-Hanning window, T=18.6 years, S=1 year)

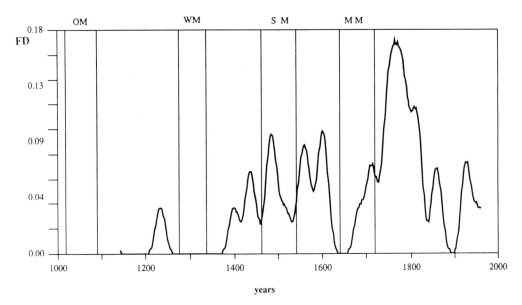

Fig. 5a Frequency distribution of the freezings of the Venetian Lagoon (Tukey-Hanning window, T=50 years, S=5 years)

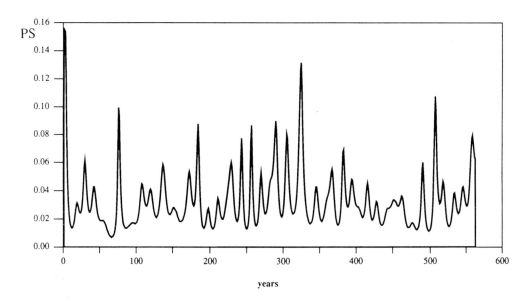

Fig. 5b Recurrence interval (yr) of the freezings of the Venetian Lagoon (MEMSA)

which was the limiting factor, preventing a severe winter being transformed into a great one. It is equally probable that this latter, latent condition will be repeated. The only period in which there were frequent great winters goes back to the end of the 1400s, that is, at the beginning of the Little Ice Age.

8. Strong winds and heavy storms

Strong winds (Fig. 6), heavy storms (Fig. 7) and hailstorms (Fig. 8) follow a similar trend, sometimes because they are phenomena associated to each other. Their (MEMSA) recurrence interval is close to the century and epoch in which gale winds and storms occurred with the maximum frequency being very close to A.D. 1500. These events reached an absolute maximum during the SM, while during the MM they were rare, as already seen with the flooding of the Adige. The time distributions of heavy rainfalls (Fig. 9) and droughts were also similar. For further detailed information about the precipitation trends during recent centuries, see other published papers (CAMUFFO, 1984; CAMUFFO et al., 1990).

9. Invasions of locusts

With reference to central-northern Italy, by comparing the various chronicles, it is possible to reconstruct the local movement of swarms of grasshoppers in areas they infested, the periods when they were more frequent, their impact on human settlements and sometimes the climatic factors (above all, steep drops in temperature) responsible for the deaths of the swarms (CAMUFFO & ENZI, 1991). The migrations followed two routes: (1) the majority of swarms started out from the Near East, travelling up the Danube basin, channelled through the Dinaric Alps and the Carpathians and, on reaching the Hungarian Plain, were transported into Italy by eastern winds (the Borino); (2) local invasions from the north were, however, very sporadic; the grasshoppers crossed the Alpine chain, descending through the Brenner Pass after having travelled along the Danube basin and then being channelled into the Inn and Adige valleys. Such invasions occurred at the same time as those coming from the east. In practice, the Apennines acted as an insurmountable barrier in that there is no mention of direct migrations of grasshoppers into northern Italy from Africa or from the Iberian peninsula (see Plinio, Nat. Hist. VI, 195; VII, 28-29).

It is believed that the migrations occurred in a succession of phases, first from the subtropical zone of origin to intermediate reception areas (suitable for a more or less long temporary stay and, also, for reproduction, such as the Pannonic Plain) and then the swarms migrated from these areas and reached Italy or central Europe, depending on the winds. The general climatic conditions in the intermediate reception areas seem to have been very important, by creating a suitable habitat for the grasshoppers to stay and reproduce.

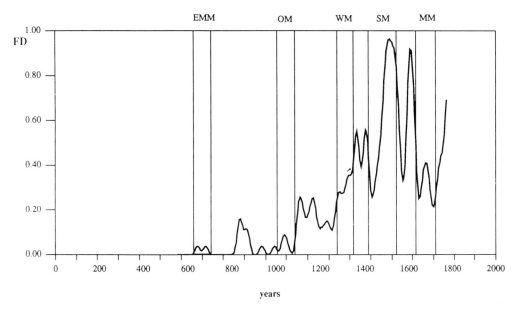

Fig. 6 Frequency distribution of the gale winds (Tukey-Hanning window, T=50 years, S=5 years)

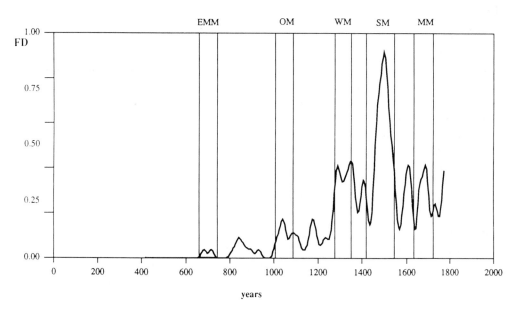

Fig. 7 Frequency distribution of the storms (Tukey-Hanning window, T=50 years, S=5 years)

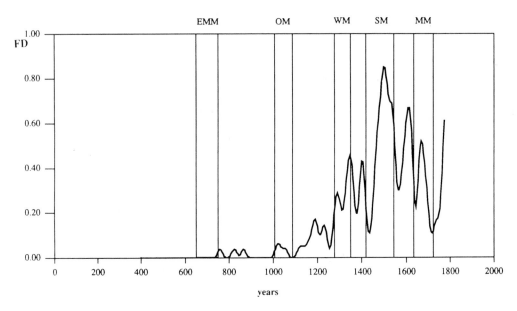

Fig. 8   Frequency distribution of the hailstorms (Tukey-Hanning window, T=50 years, S=5 years)

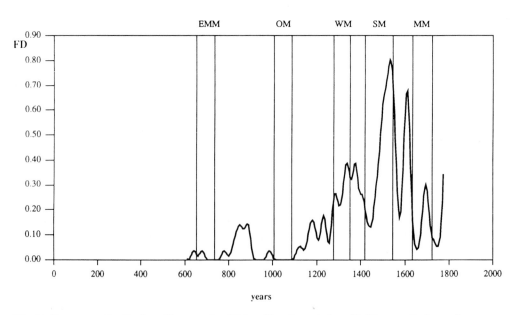

Fig. 9   Frequency distribution of heavy rains (Tukey-Hanning window, T=50 years, S=5 years)

The centuries most seriously afflicted by this calamity were the fourteenth (including the WM), the sixteenth (during the SM) and secondarily, the seventeenth (Fig. 10). The question that should be asked is, whether the great number of invasions which occurred in the fourteenth century were due, above all, to the particularly unfavourable meteorological conditions or to land being abandoned. The fact that the greater part of these invasions occurred before the great plague of 1348 would seem to indicate that the most important factor, in that particular century, was not the epidemic and the subsequent effects on the territory, but the frequency of the unfavourable meteorological conditions. It should be remembered that, notwithstanding the fact that Italy had a population of more than 5 million in the year 1000 alone, there is very little documentation about the invasions.

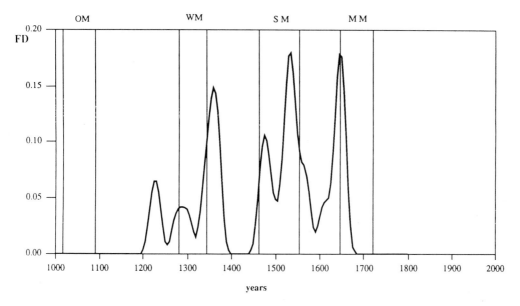

Fig. 10 Frequency distribution of the invasions of locusts (Tukey-Hanning window, T=50 years, S=5 years)

Human factors have, over the centuries, reduced the damage and the frequency of the invasions. The reduction of the frequency indicates a greater capacity of fighting the grasshoppers and a greater use of ploughing, especially in the Hungarian Plain, leading to the destruction of the eggs. This underlines the importance of agriculture in the territory, because suitable habitats can be formed in abandoned areas leading to the redevelopment of this or other similar scourges. Some areas in the Apennines, Sicily and Sardinia which are under cultivation, are still exposed to this risk.

10. Conclusions

Written sources make a precious contribution towards increasing scientific knowledge about the relationships between climatic forcing agents and the earth system's possible response. The data allow for assessing the reliability of models used to forecast future scenarios.

Among the comments reported upon, some fundamental ones should be underlined:
- The period 500-1000 A.D., traditionally held to be one of climatic deterioration should be carefully considered, since many gaps in the historical documentation occur, in particular with regard to the tenth and eleventh centuries. Therefore, it remains less well known than both the classical period which preceded it and the millenium which followed it.
- The cooling down in the Little Ice Age has demonstrated that the response of the levels of the oceans to changes in atmospheric temperature was not immediate, simple and unequivocal as some models would suggest.
- The sun-earth relationship is not easily understood and therefore cannot be easily interpreted, especially if single periods of minimum solar activity are studied separately. This complex interaction of events should be faced by inserting the various phases within the whole climatic context.
- EDDY's theory, which bases the climatic variations on the different intensity of solar emissions, should be proven on the basis of the climatic behaviour of the earth in all the periods of solar anomalies.
- From the data compiled and examined, it would seem that such a theory is not reliable. The SM was, in reality, highly tormented, but the WM and MM were not. The OM is lacking in sufficient evidence in order to draw a concrete conclusion, but in any case, the existing data do not indicate that the period was particularly troubled and they cannot, in any way, justify any such coincidence with the SM. This observation is further corroborated by the fact that, in some cases, the climatic anomalies actually preceded the solar ones. The SM occurred after a singular astronomic phase (e.g. the earth in perihelion, the moon in perigee), when the transport of ocean waters was greatly amplified by the lunar-solar forcing agents and volcanic activity was highly increased. Too little is known about the oceanic currents at that time. The possible factors forcing the climate may have been many, and are not restricted to anomalies in solar emission.

  Many phenomena of undoubted climatic interest occurred in the fifth century B.C., during the so-called Greek period minimum, and there is abundant documentation to be found in Greek literature. References from Latin literature are less frequent but equally important. At this point the problem arises of whether the Greek information can be considered really representative, in that it is known to be derived from the detailed descriptions found in Herodotus and Thucydides. That is to say, are the recurrent climatic events significant, due to a real anomaly in those periods or are they due to the wealth of literature and documentation linked to the stature of the two historians. It is a problem that is worth looking into.

-   The really exceptional phenomena, such as great winters, can occur at any time, even in periods when atmospheric heating reaches the highest levels. This means that society continues to be unprepared for natural events, when it only bases its choices on the most recent climatic trends. The current period is a particularly fortunate one which cannot, however, last much longer.

-   The anthropogenic effects on the territory could have an enormous weight, both for good and evil. For example, the invasions of grasshoppers stopped when double ploughing began to be used. This system buried the eggs very deeply in the ground, destroying them. The draining of the water-bed and the hydraulic work have had an enormously adverse effect on ground subsidence and the quantity of fresh water that is exchanged with sea water, increasing, out of all proportion, the frequency of high waters.

Acknowledgements

This article utilises the results of several studies that have been financed, on various occasions, by the European Community Commission, programme EPOCH, and by the Environmental, Territorial and Climatic Strategic Programme for southern Italy (Progetto Strategico Clima Ambiente e Territorio del Mezzogiorno), co-ordinated by the European Science Foundation. D. Camuffo is a physicist, S. Enzi a historian; both contributed with their specific competences to this interdisciplinary study.

References

CAMUFFO, D. (1984): Analysis of the series of precipitation at Padova, Italy. Climatic Change 6, 57-77

CAMUFFO, D. (1987): Freezing of the Venetian Lagoon since the ninth century A.D. in comparison to the climate of western Europe and England. Climatic Change 10, 43-66

CAMUFFO, D. (1990): Clima e Uomo. Garzanti, Milano, 207 p.

CAMUFFO, D. (1993): Analysis of the sea surges at Venice from A.D. 782 to 1990. Theor. Appl. Climatol. 47, 1-14

CAMUFFO, D. & ENZI, S. (1991): Locust invasions and climatic factors from the Middle Ages to 1800. Theor. Appl. Climatol. 43, 43-73

CAMUFFO, D. & ENZI, S. (1992): Reconstructing the climate of northern Italy from archive sources. In: Bradley, R. S. & Jones, P. D. (eds.): Climate since A.D. 1500. Routledge, London, 143-154

CAMUFFO, D.; BERNARDI, A. & ONGARO, A. (1991): Variazioni secolari delle piogge nell'Italia Settentrionale. Mem. Soc. Geogr. It. 46, 363-384

CARBOGNIN, L. & TARONI, G. (1984): Acque alte a Venezia: un'analisi statistica dei massimi annuali. Rapporti e Studi Ist. Veneto Lettere Scienze e Arti 9, 63-71

Cini Castagnoli, G.; Bonino, G. & Serio, M. (1992): Common spectral features in the 5500-year record of total carbonate in sediments and radiocarbon in tree-rings. Radiocarbon 34, 798-805

Colacino, M. & Purini, R. (1986): A study of the precipitation in Rome from 1782 to 1978. Theor. Appl. Climatol. 37, 90-96

Eddy, J. A. (1977): Climate and the changing sun. Climatic Change 1, 173-190

Lamb, H. H. (1972): Climate: Present, past and future. Methuen, London, 613 p.

Pirazzoli, P. A. (1982): Maree estreme a Venezia (periodo 1872-1891). Acqua Aria 10, 1023-1039

Rusconi, A. (1992): Le osservazioni mareografiche in Laguna di Venezia: le variazioni di livello marino osservate negli ultimi 120 anni. Bollettino Geofisico 15/5, 91-110

Wey, W. S. (1990): Time series analysis. Addison-Wesley, Redwood City/Ca, 478 p.

Authors' addresses:

Dr. D. Camuffo, CNR-ICTIMA, Corso Stati Uniti, 4, I-35020 Padova
Dr. S. Enzi, CNR-ICTIMA, Corso Stati Uniti, 4, I-35020 Padova

Variations in total solar irradiance

Claus Fröhlich

Summary

Measurements of the total solar irradiance, also called the solar constant, during the last 15 years from satellites show variations over time scales from minutes to years and decades. The most important variance is in the range from days to several months and is related to the photospheric features of solar activity: decreasing the irradiance during the appearance of sunspots, and increasing it by faculae and the bright magnetic network. Long-term modulation by the 11-year activity cycle is observed conclusively with the irradiance being higher during solar maximum. These variations can be interpreted - at least qualitatively - as manifestations of activity related features on the photosphere. Some recent findings about time dependencies of various factors limit the accuracy of any extrapolation of the present measurements to much longer time scales.

Zusammenfassung

Messungen der totalen Sonnenstrahlung, auch Solarkonstante genannt, die während der letzten 15 Jahre von Satelliten aus durchgeführt wurden, zeigen Intensitätsvariationen in Zeitskalen von Minuten, Jahren und Dezenien auf. Die wichtigste Variation, welche im Zusammenhang mit photosphärischen Eigenschaften steht, wurde im Bereich von Tagen bis zu mehreren Monaten beobachtet: wobei die Bestrahlungsstärke mit dem Erscheinen von Sonnenflecken ab- und mit Fakeln und hellen magnetischen Stürmen zunimmt. Die Langzeitmodulation des 11-Jahres Aktivitätszyklus konnte eindeutig, mit einer höheren Bestrahlungsstärke während des Sonnenmaximums, beobachtet werden. Diese Variationen können zumindest qualitativ, als Erscheinung aktiver Eigenschaften der Photosphäre zugeordnet werden. Jüngste Erkenntnisse über die zeitliche Abhängigkeit verschiedener Faktoren limitieren allerdings die Genauigkeit jeder Extrapolation bisheriger Messungen auf längere Zeiträume.

1. Introduction

The following summarizes two recent papers (FRÖHLICH, 1993, 1994) and the status of the present knowledge about possible solar forcing of the earth's climate on time scales of centuries. As the Sun is the major energy source of the earth variations of the total solar irradiance, S, may obviously influence climate, but only if the changes persist over extended periods of time because of the large thermal capacity of the oceans which strongly dampens higher frequency forcing. As the earth's surface is four times its cross-section and ~30% of the sunlight is reflected back to space without being absorbed, the mean solar heating of the earth is ~240 Wm^{-2} with $S=1367$ Wm^{-2}. A change of S by 0.1% is therefore a climate forcing of ~0.24 Wm^{-2}. The forcing by a change of S is energetically very similar to a change in the magnitude of the greenhouse effect and their effect can be directly compared: a 2% increase of S and a doubling of CO_2 correspond both to a forcing of about 4-4.8 Wm^{-2} and would yield a change of the mean earth temperature of 1.5-5.5 °C depending on the global circulation model used (see e.g. HANSEN & LACIS, 1990). The climate sensitivity is thus 0.3-1.4 °C/Wm^{-2} for a change in S or of 0.8-3.3 °C/% with a mean of about 2 °C/%.

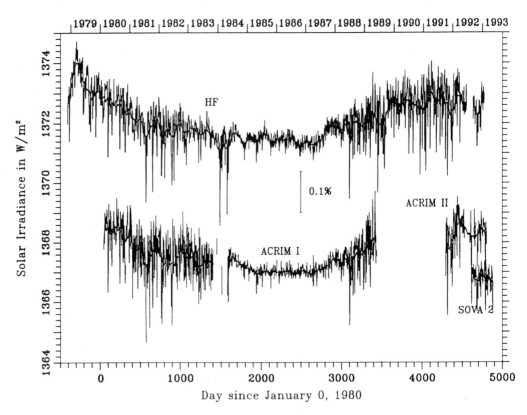

Fig. 1 Total solar irradiance as measured from satellites by NIMBUS-7 (HF), SMM (ACRIM-I), UARS (ACRIM-II) and EURECA (SOVA-2). For reference see FRÖHLICH (1994)

2. Variability on time scales up to the 11-year solar cycle

Figure 1 shows the measurements of the solar irradiance since the start of the measurements from NIMBUS 7. The dominant feature in the time series for the range of days to months is the "dip", a negative excursion of a few days' length and a depth ranging up to a few tenths of a percent of the irradiance. These dips result from large sunspot groups rotating across the visible disk. They can be described in terms of the Photometric Sunspot Index, the *PSI* function, (e.g. HUDSON & WILLSON, 1982) which is the sum of the projected areas of the sunspots multiplied by α, a factor taking into account the darkness of the sunspot relative to the photosphere, and a limb darkening function. An improved *PSI* calculation has been presented by FRÖHLICH et al. (1994). The newly calculated *PSI* was used to study its behaviour during the different phases of the solar cycles 21 and 22 by using bivariate spectral analysis for the comparison of *PSI* and irradiance time series. The result shows a strong time dependence of the gain, the factor by which *PSI* has to be multiplied to yield the observed irradiance change. It changes from about 0.6 in 1980 to 1.1 in 1990 and seems to decline after that time. An obvious reason for a change could be the contrast used in the *PSI* calculation. As the irradiance is an average over the disk, the observed "contrast" is between the spots and their surrounding in a disk-averaged sense and not as a temperature difference between the umbra and penumbra of the sunspots and the quiet Sun. Thus, it is more likely to be a change of relative area of spots and bright features, the faculae and the bright network surrounding the spots in the active region. The fact that this time dependence seems to be unrelated to the 11-year sunspot cycle, obviously questions extrapolations to other cycles. Independent of its interpretation, this time-dependent factor is a kind of "calibration" of *PSI*, which can be utilized to withdraw the effect of sunspots from the irradiance observations. Figure 2 shows the 81-day running mean values of the reduced irradiance (*S+PSI*); the amplitude of the cycle variation is increased, thus the enhancement of the radiation during high activity overwhelms the reduction by spots. Part of the enhancement may be hidden in the *PSI* as temporal variation of the contrast.

WILLSON (1982) called attention to the effects of faculae in modulating the irradiance and having the potential to compensate part of the sunspot deficit and FOUKAL & LEAN (1988) and WILLSON & HUDSON (1988) have shown that faculae and the bright network can also explain the cycle variation. Direct photometric approaches have yielded some insight into the relationship between the irradiance excesses due to faculae of an active region and the sunspot deficits (e.g. CHAPMAN, 1987; FOUKAL, 1990; LEAN, 1991). The area and contrast of faculae, however, are difficult to observe and are not as well known as those of the spots. Thus a facular photometric index - similar to *PSI* - is difficult to establish. Moreover, such a facular index would neglect the influence of the bright network. Thus, proxy data seem to be the better, although less physical, approach. The equivalent width of the He 1083 nm line as measured for the full disk, the He I index, is known to be representative of the magnetic elements, including both active-region and network facular elements (HARVEY, 1984) and FOUKAL & LEAN (1988) have identified this index as suitable proxy for the influence of magnetic elements on total irradiance. The relation between both is determined by a simple linear regression analysis. This correlation was quite good for the declining phase of cycle 21, but it does not hold for the full period. If the linear regression between He I and irradiance is performed over the full period it fails to explain the two maxima simultaneously: it overestimates the maximum of 21 and underestimates the one of 22. In order to get more insight in the temporal behaviour of this relationship FRÖHLICH (1994) divided the full period into intervals of 1-2 years length (as it was done for *PSI* by FRÖHLICH et al., 1994) and calculated the linear regression for each interval yielding a time sequence of intercepts (irradiance at zero He I) and slopes. A linear trend and a sine wave with 11-year period can be fitted to the slopes and intercepts respectively, yielding a maximum of the slope in 1988 and one for the intercept in 1992. Moreover, the slope has a positive trend producing a behaviour similar to the contrast in the *PSI* representation of the spot influence: it is low in 1980/81, increasing until 1988 and slightly decreasing towards 1992. The intercept does the opposite and weakens the slope effect to some extent. As we are dealing with a factor of 3.5 between minimum and maximum slope, however, the weakening due to the out-of-phase change of the intercept is not very much influencing the net effect. Part of this time dependence could arise from the fact that some of the facular influence was already taken into account by the varying contrast of *PSI*. The irradiance reconstructed using the fitted slope and intercept is shown in Fig. 2. The fit is quite good although some discrepancy seem to be too large to be accounted for by uncertainties in the index or irradiance. Again, as with *PSI*, the time dependence of the slope and intercept of the He I model makes any extrapolation to other time periods difficult.

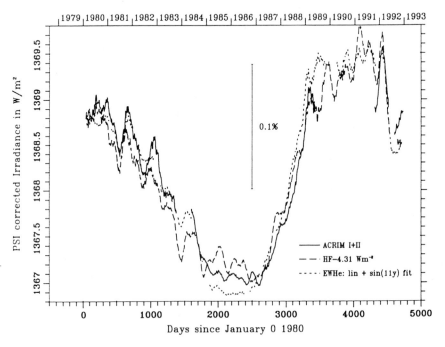

Fig. 2 Solar irradiance measured by ACRIM and HF with the effect of the sunspots removed (*S+PSI*). The short dashed line represents a model based on the He I index calculated with the time-dependent slopes and intercepts (see text)

3. Possible variations on longer time scales

EDDY (1976) has argued that reduced solar irradiance during the Maunder Minimum of solar activity (1640-1720) may have been responsible for the coldest period of the Little Ice Age when global temperature is estimated to have been as much as 1°C colder than today (e.g. WIGLEY & KELLY, 1990). Although the direction (low irradiance during low activity) is correct, the observed amplitude of the 11-year modulation of ~0.1% is about 5 times too small (see Introduction). Moreover, reconstructions of S, back to 1884 (FOUKAL & LEAN, 1990), using PSI and the strong correlation between sunspot numbers and F10.7, show that solar cycle 21 and 22 amplitude was the largest in the whole series on the basis of sunspot and facular effects. During solar minimum, however, a surface network of bright magnetic elements remains on the Sun (e.g. FOUKAL et al., 1991). If this had been removed during a Maunder Minimum period an even lower irradiance would have been observed. From the model proposed by FOUKAL et al. (1991) the irradiance would be lowered by 1.8 Wm^{-2} if the entire network were completely removed. The presence of bright magnetic elements is well indicated by e.g. the emission reversal in the core of the chromospheric CaII H and K absorption lines. H and K fluxes are also observed from stars as measure of stellar activity. The HK index is defined as the ratio of the flux in narrow pass-bands centred on the H and K emission cores compared to the flux in the nearby continuum, and varied for the Sun between 0.174 during the solar maximum in 1980/82 and 0.166 during the minimum in 1985/87 (LIVINGSTON & WHITE in GIAMPAPA, 1990). More recent work (LEAN et al., 1992) on the translation of solar CaII observations to the HK index indicate that the last solar cycle had more likely a variation between 0.175 and 0.195 (from Fig. 1 of LEAN et al., 1992; in the following these values are given in parenthesis for comparison). Investigation of the activity of solar type stars (BALIUNAS & JASTROW, 1990) indicate that there is a group of stars which exhibit Maunder Minimum type "non-activities" and have HK indices of 0.146 or 2.5 (1.5) times the solar cycle 21 amplitude lower than its solar minimum value. From Fig. 2 the "magnetic" irradiance amplitude amounts to about 2 Wm^{-2} for cycle 21 or the Maunder Minimum value would have been 5.25 (3.05) Wm^{-2} below the minimum in 1985/87 if a linear relationship is assumed and if the seventeenth century Sun behaved like the low activity reference stars. The linear relationship is somewhat questioned by the time-dependent slope of the He I-irradiance regression and makes this extrapolation uncertain. Nevertheless, it would mean a 0.38% (0.22%) lower irradiance (below the minimum between cycle 21 and 22) and is about double the value expected from only removing the network, which indicates that for a non-cycling Sun some non-network magnetic fields would have to be removed also as stated by LEAN et al. (1992). As already mentioned by BALLIUNAS & JASTROW (1990) this might "explain" part of the lower temperatures during the Little Ice Age by activity related changes of the solar irradiance. The uncertainties are such that the values stated have to be cautioned; the irradiance, however, was lower during the Maunder Minimum, the question is how much. On the other hand the estimates for global temperature changes by GCM are based on direct energetic forcing - maybe there are amplification mechanisms for the much stronger solar UV-irradiance changes through stratosphere-troposphere couplings.

References

BALIUNAS, S. & JASTROW, R. (1990): Evidence for long-term brightness changes of solar-type stars. Nature 348, 520-523

CHAPMAN, G. A. (1987): Variations of solar irradiance due to magnetic activity. Ann. Rev. Astron. Astrophys. 25, 633-667

EDDY, J. A. (1976): The Maunder Minimum. Science 192, 1189-1202

FOUKAL, P. V. (1990): Solar luminosity over time scales of days to the past few solar cycles. Phil. Trans. Roy. Soc. London A 330, 591-599

FOUKAL, P. V. & LEAN, J. (1988): Magnetic modulation of solar luminosity by photospheric activity. Astrophys. J. 328, 347-357

FOUKAL, P. V. & LEAN, J. (1990): An empirical model of total solar irradiance variation between 1874 and 1988. Science 247, 556-558

FOUKAL, P. V.; HARVEY, K. & HILL, F. (1991): Do changes in the photosperic network cause the 11-year variation of total solar irradiance? Astrophys. J. 383, L89-L92

FRÖHLICH, C. (1993): Changes of total solar irradiance. In: Interactions between global climate subsystems, The Legacy of Hann, Geophys. Monogr. 75, IUGG 15, 123-129

FRÖHLICH, C. (1994): Irradiance observations of the sun. In: Pap, J.; Fröhlich, C.; Hudson, H. S. & Solanki, S. (eds.): The sun as a variable star, solar and stellar irradiance variations. Cambridge Univ. Press, 28-36

FRÖHLICH, C.; PAP, J. & HUDSON, H. S. (1994): Improvement of the photospheric sunspot index and changes of the effective sunspot influence with time. Sol. Phys. 152, 111-118

GIAMPAPA, M. S. (1990): The solar - stellar connection. Nature 348, 488-489

HANSEN, J. E. & LACIS, A. A. (1990): Sun and dust *versus* greenhouse gases: an assessment of their relative roles in global climate change. Nature 346, 713-719

HARVEY, J. (1984): Helium 10830 Å Irradiance: 1975-1983. In: LaBonte, B. J.; Chapman, G. A.; Hudson, H. S. & Willson, R. C. (eds.): Solar irradiance variations on active region time scales. NASA Conf. Publ. CP-2310, 197-212

HUDSON, H. S. & WILLSON, R. C. (1982): Sunspots and solar variability. In: Cram, L. & Thomas, J. (eds.): Physics of sunspots. Sacramento Peak Observatory, 434-445

LEAN, J. (1991): Variations in the sun's radiatiove output. Rev. Geophys. 29, 505-535

LEAN, J.; SKUMANICH, A. & WHITE, O. (1992): Estimating the sun's output during the Maunder Minimum. Geophys. Res. Lett. 19, 1591-1594

WIGLEY, T. M. & KELLY, P. M. (1990): Holocene climatic change, ^{14}C wiggles and variations in solar irradiance. Phil. Trans. Roy. Soc. A330, 547-560

WILLSON, R. C. (1982): Solar irradiance variations and solar activity. J. Geophys. Res. 87, 4319-4324

WILLSON, R. C. & HUDSON, H. S. (1988): Solar luminosity variations in solar cycle 21. Nature 332, 332-334

Author's address:

Dr. C. Fröhlich, Physikalisch-Meteorologisches Observatorium Davos, World Radiation Center, CH-7260 Davos Dorf

The Maunder Minimum and the deepest phase of the Little Ice Age: a causal relationship or a coincidence?

Elizabeth Nesme-Ribes

Summary

There is little doubt that the increase of greenhouse gases in the earth's atmosphere will affect our future climate. However, it is clear that other parameters in the past have had their share in creating climate fluctuations observed over the last millennia. Solar variability could be one of these parameters. During the second half of the seventeenth century, very few sunspots were sighted. This period, known as the Maunder Minimum, coincided with a long and severe cold on earth, called the Little Ice Age. In this paper, we shall describe the manifestations of the Maunder Minimum, and estimate the solar output during that time. The main characteristics of the "Little Ice Age" will be described in this issue. The impact of the Maunder Minimum on the earth's climate will be addressed by R. Sadourny.

Résumé

S'il ne fait aucun doute que l'augmentation de la teneur des gaz à effet de serre dans l'atmosphère terrestre risque de modifier le climat futur, il est clair que d'autres paramètres ont pu participer aux variations climatiques observées au cours des derniers millènaires. La variabilité solaire pourrait être l'un d'eux. Pour preuves, au 17e siècle, le Soleil a connu une anomalie, le Minimum de Maunder, laquelle a coincidé avec une période extrèmement froide, le Petit Age Glaciaire. Dans cet article, nous montrerons comment cette anomalie solaire s'est manifestée et nous examinerons les conséquences possibles en termes de l'énergie rayonnée sur terre. Les caractéristiques du Petit Age Glaciaire seront abordées dans d'autres articles, et la réponse du forçage solaire sur le climat terrestre fera l'objet de la présentation de R. Sadourny.

1. Introduction

When investigating the possible relationship between the Maunder Minimum and the "Little Ice Age", three phases are required. The first phase concerns the existence of a long lull of solar activity during the seventeenth century. The decrease of magnetic activity has long been questioned as it was not clearly established whether the solar anomaly was real or was rather due to a spotty observations of the sun. Furthermore, one has to estimate the amplitude of the solar anomaly in terms of solar energy output. Both points will be discussed in Section 2.

The second phase has to do with the "Little Ice Age", which began in the sixteenth, and lasted until the eighteenth century, with a deep cold at the end of the seventeenth century. Was this phenomenon global in its extent, or was it an event mainly restricted to western Europe and the North Atlantic area? To address this question, one has to rely on palaeoclimatic data from a widely distributed number of points on the planet. This point shall be covered by some of the papers invited for presentation here (see M. Hughes for example).

Next, assuming the first two phenomena were real and coincident, was there any causal relationship between them? For that, we need to understand the cause and the amplitude of the solar forcing. This aspect will be discussed in Section 2. Lastly, the climate response of the solar forcing should be compatible with the available palaeoclimatic data. This will be covered in the following paper (R. Sadourny).

2. The Maunder Minimum: its properties

The difficulty in explaining what occurred during the Maunder Minimum arises from the fact that France was the only country where a systematic observation of solar activity was performed over the period 1650 to 1719. In several European countries, astronomers were very active, but after the excitement of the discoveries of sunspots and the sun's rotation, these other brilliant astronomers switched to other celestial objects. The French astronomers decided to make a systematic survey of the sun to address a number of problems of physics: atmospheric refraction, accurate determination of the orbital parameters of the earth-sun system, the exact position of fixed stars in a reference system, mapping of France and distant countries, and absolute time measurements. More than 8000 days of solar observations covering 70 years were accumulated by CASSINI (1729), PICARD (1671) and LA HIRE (1729) among others. Their results have been analysed by RIBES & NESME-RIBES (1993) and can be summarized as follows.

2.1 The solar activity: sunspots, faculae, corona

The sunspot activity level was extremely low from 1660 to 1704, and sunspots were confined to the southern hemisphere only. The latitudes covered by those southern sunspots hardly exceeded 10%. Then, activity slowly resumed in both hemispheres from 1714 to 1719, with sunspots covering a larger latitude range (up to 20%). Few faculae were detected, in contrast with those discovered by SCHEINER (1630) and described at length in his "Rosa Ursinae".

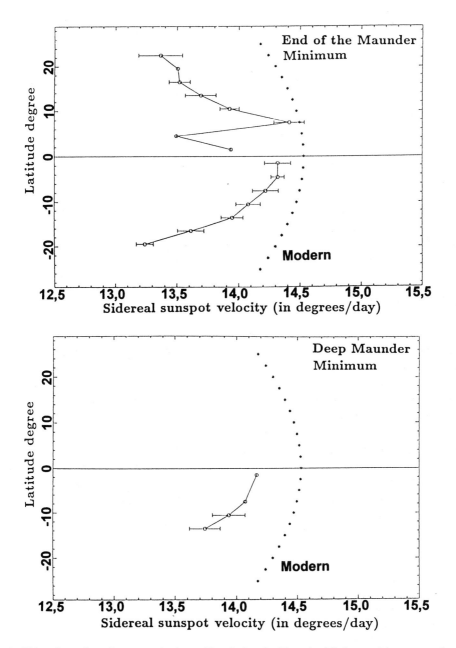

Fig. 1 Sidereal rotation of sunspots in degree/day during the Maunder Minimum. (a) corresponds to the deep phase (1660-1705), (b) to the resuming period (1711-1719). Sidereal rotation rate for modern sunspots (asterisks) is shown for comparison

2.2 Solar rotation

The solar rotation at that time could be reconstructed. Sunspots, when observed, were positioned in latitude and longitude with respect to the solar disk. This way it was possible to reconstruct their motions across the solar disk (RIBES et al., 1987). The rotation of sunspots during the period of very low activity (from 1660 to 1704, referred hereafter as "the deep Maunder Minimum") was slower by several per cent, with respect to the modern rate (Fig. 1). The latitude of sunspots at 10° was found to be the same as the present rate of the internal radiative zone obtained from seismic measurements (RIBES et al., 1992) (Fig. 2a). At the end of the Maunder Minimum, from 1705 to 1719, sunspot rotation slowed down further (Fig. 1b), with latitudes of 20° rotating at the speed of the radiative interior (Fig. 2b). For comparison, internal rotation deduced from seismic observations is shown in Fig. 2c.

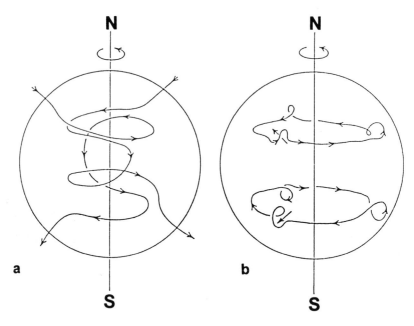

Fig. 2 Schematic representation of the α-ω dynamo mechanism. A non-uniform rotation distorts the general dipole and creates an azimuthal (east-west) oriented magnetic line antisymmetric with respect to the solar equator. Cyclonic motions restore the general dipole field

2.3 Apparent solar diameter

The solar disk also exhibited an apparent expansion of ~2.10⁻³ during the deep Maunder Minimum. There are several (non solar) reasons which could account for an apparent expansion of the solar diameter: poor quality of the seventeenth century optics and special terrestrial atmospheric conditions would enlarge the solar image. However, a careful investigation of the various sources of error led the author to think that this expansion was real (RIBES et al., 1989; RIBES et al., 1992). Furthermore, the variation in the solar diameter was accompanied by a 10-year modulation which is negatively correlated with sunspot activity. A similar modulation is present in cosmogenic isotope abundances. Both modulations indicate that the solar cycle was still operating at the bottom of the convective zone, in spite of the reduced surface manifestation (RIBES et al., 1989; KOCHAROV, 1986; STUIVER & BRAZIUNAS, 1993).

2.4 Coronal activity and aurorae

The state of the corona was reported by the French observers at the time of solar eclipses. Their reports indicate that solar activity was low (RIBES & NESME-RIBES, 1993). This again finds some support in the aurorae observations in Europe at that time (LINK, 1977).

3. Solar output during the Maunder Minimum

It is clear that we have no direct measurement of solar output during the Maunder Minimum, as total solar irradiance has been monitored only since 1978 (HICKEY et al., 1988; WILLSON & HUDSON, 1991). So we have to rely on surrogates to estimate the variation of the solar output. What are these surrogates?

3.1 Solar activity: sunspots and faculae

There is a striking correlation between the sunspot number and the variation of the total solar irradiance on an 11-year time scale (FOUKAL & LEAN, 1988). This was surprising at first, because sunspots are cooler than the surrounding atmosphere and should thereby reduce the total solar energy output. This is indeed observed on a time scale of a few days: an abrupt decrease (up to several per thousands) in the total solar irradiance occurs whenever sunspots transit across the solar disc.

Conversely, faculae lead to some short-term increases of the total solar irradiance. It was conjectured that sunspots and faculae might balance each other, so that the total solar irradiance would be constant over an 11-year cycle. This is certainly not the case, as has been seen in the measurements of the "solar constant". Some other effects should contribute to the total solar irradiance in order to account for the observed variation (PAP & FRÖHLICH, 1992; WILLSON & HUDSON, 1991).

Another emissive feature consists of bright points denoted hereafter as "network". These are probably the seat of strong, isolated magnetic fields. It is not clear what they contribute to the total solar irradiance and how they vary through the solar cycle. It has been conjectured that all traces of activity were gone during the Maunder Minimum. By removing all surface activity, i.e. sunspots, faculae and network, LEAN et al. (1992) have given some estimate for the decrease of the total solar irradiance (~.2%). This is not substantiated however as there is no evidence that activity manifestation other than sunspots was absent. On the other hand, small-scale or large-scale motions could also influence the total solar irradiance. This point will be addressed in the next subsection.

3.2 Solar rotation

Many of the above observations can be explained on the basis of our present knowledge of the modern 11-year solar cycles. Let us first summarize what we know about the modern 11-year solar cycles and establish how the solar dynamo is thought to operate. PARKER (1988) imagined that non-uniform rotation (in the radial direction) could be the source of azimuthal fields (that of active regions). A general magnetic field exists in the form of a dipole oriented towards the axis of rotation and a non-uniform rotation could distort the poloidal magnetic lines and generate some east-west oriented magnetic lines (Fig. 3). These azimuthal fields would emerge at the surface in the form of small (e-w) magnets, each pole of which is a sunspot or a facula of a given polarity. Solar rotation is the first key ingredient to generate active regions. A second key ingredient is necessary to make this dynamo "cyclic". Turbulence and some amount of large-scale convection have been suggested as providing the helicity required to restore the general dipole magnetic field. Solar activity is thus the surface manifestation of a dynamo wave resulting from the combination of non-uniform rotation and certain non-axisymmetric motions.

An additional constraint has been put on the direction of propagation of this dynamo wave. We know from observation that solar activity starts at high latitudes at the beginning of a new cycle and moves equatorwards as the cycle progresses. To introduce this direction of propagation into the dynamo equations, the radial shear $d\omega/dr$ in the angular velocity must be negative, i.e. the velocity should increase with depth, assuming that helicity is positive in sign in the northern hemisphere. Then, a negative dynamo number (proportional to $d\omega/dr$ times α) corresponds to an equatorward dynamo wave and accounts for the observed trend in the distribution of solar activity, the so-called "butterfly diagram" (Fig. 4a).

However, recent seismic observations indicate how the angular velocity varies from the surface down to .4 R_o. One very interesting point is that the negative shear is in fact restricted to the surface latitudes from 35 to 90°, while shear is positive at latitudes ranging from the equator to 35° (Fig. 2a). These results are puzzling, as we expected the assumed negative shear near the equator to account for the branch of sunspot activity (the "butterfly diagram"), see Fig. 4. To reconcile theory and observations, either the helicity (which is a parameter difficult to observe) has a sign opposite to what was expected, and an equatorward wave is still produced in the region where the shear is positive (GLATZMAIER, 1985), or there are two dynamo waves interacting, one poleward and one equatorward. The main wave is generated in the region where the dynamo number is large (around 70°), and propagates equatorwards because the shear is negative. A poleward branch is possibly generated near the equator, with a small dynamo number (because the helicity near the equator is small). Evidence for this secondary dynamo wave is quite new (MOURADIAN & SORU-ESCAUT, 1993), so the interaction of the dynamo waves has not yet been discussed theoretically.

The dynamo theory is still conjectural (PARKER, 1988). However, one very interesting point of Fig. 2 is that the surface rotation reflects the amplitude of the internal shear of the angular velocity, which is the reservoir of azimuthal fields (NESME-RIBES, 1990). So we can relate the solar rotation of sunspots during the Maunder Minimum, and also in modern cycles, to the importance of the radial shear and thereby to the strength of the solar cycle. In particular, the decrease of solar rotation during the seventeenth century provides additional support for the low sunspot activity. It is premature to speculate on the two branches of activity. However, the positive radial shear was small, which is consistent with the latitudes of sunspots never restricted to latitudes below 10 or 20° (Fig. 4a).

Can we go a step further, i.e. can we relate the change of solar rotation to the change of solar output? This is a difficult problem to solve in its generality, so we shall use simple models to provide some orders of magnitude. Assuming that the thermo-nuclear reactions do not change over time scales of centuries, the energy flux coming from the solar interior is supposed to be constant. This energy can be stored in various ways, however, partly in the form of magnetic energy, partly in the form of kinetic energy (rotation), partly in the form of gravitational (diameter) and thermal energy (luminosity). The solar cycle is simply a change in the content of each energy reservoir (RIBES & LACLARE, 1988). One simple way of estimating the thermal energy reservoir is to assume that large-scale motions, such as these detected at Meudon (RIBES & BONNEFOND, 1990; NESME-RIBES et al., 1993b) are the third component that could explain the 11-year trend in the total solar irradiance.

The meridional circulation found at Meudon is strongly time-dependent. The number and the velocity of the vortices change with the phase of the cycle. We have therefore examined the possibility that these might account for the observed 11-year trend. Within the limitations of the theoretical model, the answer was rather interesting: namely, the kinetic energy associated with the meridional circulation was greater at the time of sunspot minimum, for the modern 11-year cycles. Both the excess kinetic energy at the sunspot minimum and the decrease of the total solar irradiance have similar amplitudes and opposite signs. So we speculated that the excess kinetic energy was borrowed from the thermal energy reservoir, thereby explaining the observed decrease of the total solar irradiance (NESME-RIBES et al., 1993a).

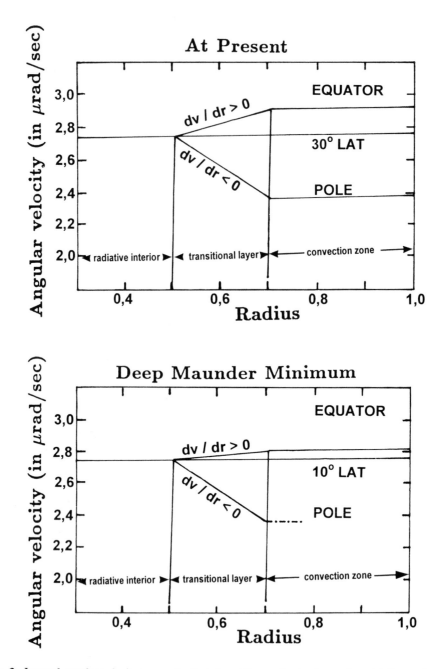

Fig. 3 Internal angular velocity *versus* depth deduced from seismic observations (a). Internal angular velocity *versus* depth has been reconstructed from sunspots observed during the deep Maunder Minimum (b) and in the resuming period of solar activity (c)

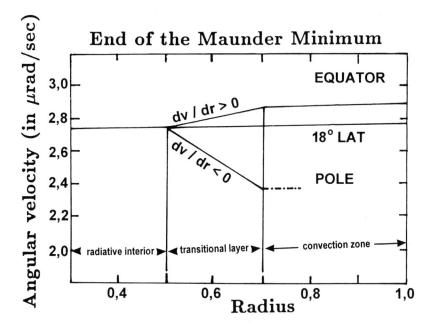

Fig. 3 continued

We developed this idea further by exploring the characteristics of the meridional circulation and rotation during the Maunder Minimum (NESME-RIBES & MANGENEY, 1992) and found that observations were compatible with a very efficient meridional circulation, carrying a strong kinetic energy. This energy had to come from some reservoir, possibly the thermal energy reservoir. The excess kinetic energy was four times larger during the Maunder Minimum than during the present solar cycle, implying that the total solar irradiance decreased by .2% at time of sunspot activity. We conclude that the solar dynamics (rotation and meridional circulation) can be used as a surrogate for the total solar irradiance. A limitation to this, however, is that the surface rotation, during the Maunder Minimum, was deduced from the sunspot motions. When there was no sunspot activity, there is no way in estimating the solar rotation. Other proxy data are necessary. One of them is the solar diameter.

3.3 Solar diameter

There is a connection between the changes in the radius and in the luminosity of the sun (see SPRUIT, 1992 for a review of the mechanisms). The problem is that the physical processes involved depend upon the time scales. Most of the magnetic field effects concern the

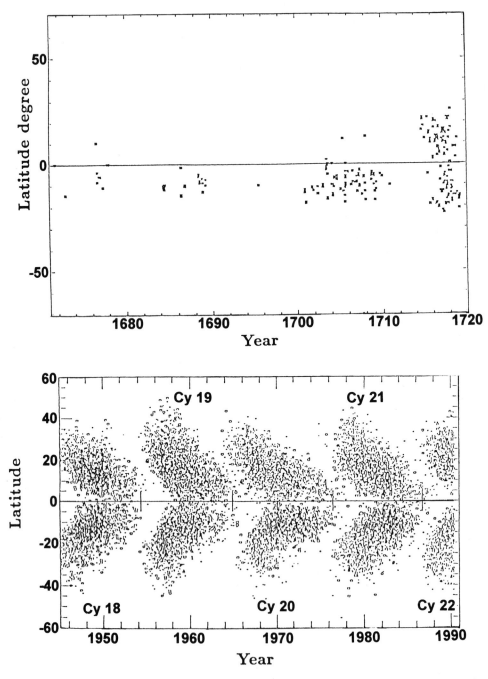

Fig. 4 Butterfly sunspot diagram during the Maunder Minimum, reconstructed from observations by the French School of Astronomy

change in the convective energy transport (SPRUIT, 1992). For 11-year time scales, the model of SPIEGEL & WEISS (1980) leads to a negative correlation between the changes in the radius R and in the luminosity L, so W = dLnR / R /dLn L / L is negative. This is in rough agreement with the observations of W = -0.2 for the latest 11-year solar cycle.

Applying the same relationship during the Maunder Minimum, which is indeed empirical, we can deduce the total solar irradiance from the seventeenth century radius observations. The result is shown in Fig. 5. It is clear that a modulation exists, the total solar irradiance being larger at times of relative sunspot activity, as it does now. Moreover, the maximum total solar irradiance is compatible with the estimate obtained from other surrogates, such as sunspot rotation. The change in output varies from .1% to 1%, with a mean value of 0.4%. This corresponds to a mean solar forcing of ~1 Wm^{-2} at the top of the earth's atmosphere, which is large enough to cause some fluctuation in the climatic models (NESME-RIBES et al., 1993a). Detailed results will be given by SADOURNY (this volume).

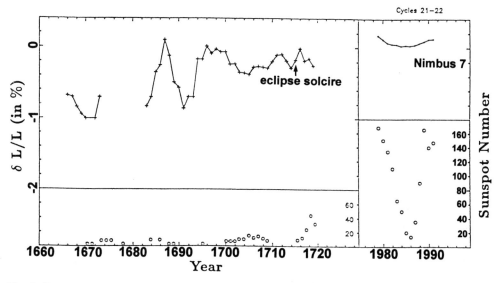

Fig. 5 Reconstruction of the total solar irradiance at the time of the Maunder Minimum (upper left). Two independent solar indicators (sunspot rotation and solar diameter) have been used as surrogates for the total solar irradiance. Sunspot number is plotted (inner left). Similar plots are drawn for the modern cycle (upper and inner right)

4. Star brightness

As shown in other contributions (i.e., STUIVER, this volume), there are some indications that the sun has experienced several "anomalies" over millenia. One crucial question is the amplitude of the solar forcing associated with these episodes. Right now, we cannot give any figures. However, there is an interesting approach developed by BALIUNAS and collaborators, using the brightness variations of solar-type stars (BALIUNAS & JASTROW, 1990). We know that solar-type stars probably exhibit magnetic cycles similar to solar cycles, although there is no straightforward evidence for this. This is because magnetic fields are extremely difficult to resolve on stars (SEMEL, 1989). On the other hand, these stars do show brightness cycle variations, the cause of which is attributed to magnetic fields. In particular, the CaII line emission is a good surrogate for the magnetic activity (LEAN et al., 1992). By comparison with the sun observed as a star, LEAN et al. (1992) have shown the existence of two populations of solar-type stars, one having a CaII emission more intense than the other. The sun lies near the border between the two. This is an indication that the sun is very close to a state of low magnetic activity, in which excursions to Maunder Minimum states could take place, with a possible decrease of the brightness. Using the variation of star brightness, it is also possible to infer how much the solar output can vary, as shown by FRÖHLICH (this volume). The estimate of total solar irradiance decrease deduced from knowledge of stars is similar to ours, within the given uncertainties.

5. Conclusion

On the astronomical side, a better understanding of the solar dynamo is needed as well as of its consequences in terms of transfer of energy in the various reservoirs. The rotation of magnetic tracers should be useful in analyzing seismic observations of the internal rotation, and thus in determining the role of the magnetic field. The study of solar-type stars is a promising field for improving our knowledge of solar activity. Long minima or maxima of solar (stellar) activity cycles occur periodically. They can be tracked for the sun by using the abundance of the cosmogenic isotopes. It is essential to identify these and estimate their consequences in terms of solar output. A lot of modelling remains to be done for interpreting these abundances in terms of solar magnetic field intensity, shape of the dipole magnetic field, and so on. Observational dating problems are also of importance.

On the climate side, palaeoclimatic data over recent centuries are crucial to establish the reality of a world wide "Little Ice Age". If this cooling during the sixteenth and the seventeenth centuries is related to a decrease of solar radiation output, climatic models do predict a strong response at both the earth's poles. Monsoon rainfalls should also be very sensitive to a solar signal. As pointed out by SADOURNY (this volume), there is a pressing need to collect such palaeoclimatic data or historical reports for a better understanding of the possible climate forcing.

Acknowledgments

This work has been supported by military contract DRET N° 922011/A. To interpret historical solar cycles, modern 11-year cycles must be properly understood. The prerequisite accuracy has been achieved on modern spectroheliograms by means of the Machine à Mesurer pour l'Astronomie (MAMA) of the Institut National des Sciences de l'Univers at the Paris Observatory. My thanks go to the MAMA team as well as to the observers who carry out the systematic observations at the Paris Observatory, in Meudon.

References

BALIUNAS, S. & JASTROW, R. (1990): Evidence for long-term brightness changes of solar-type stars. Nature 348, 520-523

CASSINI, J. D. (1729): Anc. Acad. Sci. Paris Tome VIII, 82

CHAPMAN, G. (1992): Variations of solar irradiance due to magnetic activity. Ann. Rev. Astron. Astrophys. 25, 633-667

EDDY, J. (1976): The Maunder Minimum. Science 192, 1189-1202

FOUKAL, P. & LEAN, P. (1988): Magnetic modulation of solar luminosity by photospheric activity. Astrophys. J. 328, 347-357

FRÖHLICH, C. (1995): Variations in total solar irradiance. In: Frenzel, B.; Nanni, T.; Galli, M. & Gläser, B. (eds.): Solar output and climate during the Holocene. Paläoklimaforschung/ Palaeoclimate Research 16 (this volume), 125-130

FRÖHLICH, C. & PAP, J. (1988): Multi-spectral analysis of total solar irradiance variations. Astronaut. Aeron. 220, 272-280

GLATZMAIER, G. (1985): Numerical simulations of stellar convective dynamics, Vol. I: the model and the method. J. Comput. Phys. 55, 461-484

HICKEY, J. R.; ALTON, B. M.; KYLE, H. L. & HOYT, D. V. (1988): Total solar irradiance measurements by ERB/NIMBUS 7: a review of nine years. Space Sci. Rev. 48, 321-342

KOCHAROV, G. E. (1986): Cosmic ray archaeology, solar activity and supernova explosions. In: ESA SP-25 (ed.): Proceedings of the joint Varenna-Abustumani International School and Workshop on Plasma Astrophysics, Sukhumi, CIS, 259-270

LA HIRE, P. (1729): Archives Observ. Paris, manuscripts D 2, 1-10 (1683-1719)

LEAN, J.; SKUMANICH, A. & WHITE, O. R. (1992): Estimating the total solar irradiance during the Maunder Minimum. Geophys. Res. Lett. 19/15, 1591-1594

LINK, F. (1977): Sur l'activité solaire au 17e siècle. Astronaut. Aeron. 54, 857-861

MOURADIAN, Z. & SORU-ESCAUT, I. (1993): On solar activity and the solar cycle. A new analysis of the butterfly diagram of sunspots. Astron. Astrophys. 280, 661-665

NESME-RIBES, E. (1990): Longs cycles d'activité solaire. In: Benest, D. & Froeschlé, C. (eds.): Le Soleil, une étoile et son domaine. Ecole de Goutelas, 357-373

NESME-RIBES, E. & MANGENEY, A. (1992): On a plausible physical mechanism connecting the Maunder Minimum to the Little Ice Age. Radiocarbon 34/2, 263-270

NESME-RIBES, E.; FERREIRA, E. N. & MEIN, P. (1993b): Solar dynamics over cycle 21 using sunspot as tracers. I. Sunspot rotation. Astron. Astrophys. 274, 563-570

NESME-RIBES, E.; FERREIRA, E. N.; SADOURNY, R.; LE TREUT, H. & LI, L. (1993a): Solar dynamics and its impact on solar irradiance and the terrestrial climate. J. Geophys. Res. 98, A 11, 18923-18935

PAP, J. & FRÖHLICH, C. (1992): Multi-variate spectral analysis of short-term irradiance variations. In: Donelly, R. F. (ed.): Proc. of the Workshop on the Solar Electromagnetic Radiation Study for Solar Cycle 22, 62-73

PARKER, E. N. (1988): The dynamo dilemma. Sol. Phys. 110, 11-21

PICARD, J. (1671). In: Arch. Obs. Paris, manuscripts D 1, 14

RIBES, E. & BONNEFOND, F. (1990): Magnetic tracers, a probem of the solar convective layers. GAFD 55, 241-245

RIBES, E. & LACLARE, F. (1988): Toroidal convective rolls: a challenge to theory. Geophys. Astrophys. Fluid Dynamics 41, 171-180

RIBES, J.-C. & NESME-RIBES, E. (1993): The solar sunspot cycle in the Maunder Minimum A.D. 1645 to A.D. 1715. Astron. Astrophys. 276, 549-563

RIBES, E.; BEARDSLEY, B.; BROWN, T. M.; DELACHE, P.; LACLARE, F.; KUHN, J. & LEISTER, N. V. (1992): Variability of the solar diameter. In: Giampapa, M. & Matthews, M. (eds.): The Sun in Time, 59-97

RIBES, E; MEIN, P. & MANGENEY, A. (1985): A large-scale circulation in the solar convective zone. Nature 318, 170-171

RIBES, E.; RIBES, J.-C. & BARTHALOT, R. (1987): Evidence for a larger sun with a slower rotation. Nature 326, 52-55

RIBES, E.; RIBES, J.-C. & BARTHALOT, R. (1988): Size of the sun in the seventeenth century. Nature 332, 689-690

RIBES, E.; RIBES, J.-C.; MERLIN, P. & BARTHALOT, R. (1989): Absolute periodicities in the solar diameter, derived from historical and modern measurements. Annales Geophysicae 7, 321-329

SADOURNY, R. (1995): Climate sensitivity, Maunder Minimum and the Little Ice Age. In: Frenzel, B.; Nanni, T.; Galli, M. & Gläser, B. (eds.): Solar output and climate during the Holocene Paläoklimaforschung/Palaeoclimate Research 16 (this volume), 145-159

SCHEINER, C. (1630) Rosa Ursina sive sol. Edited by Andreas Phaeus Braccuabi

SEMEL, M. (1989): Zeeman-Doppler imaging of active stars. I. Basic principles. Astronaut. Aeron. 225, 456-466

SPIEGEL, E. A. & WEISS, N. O. (1980): Magnetic activity and variations in solar luminosity. Nature 287, 616-617

SPRUIT, H. (1992): Luminosity and radius variations. In: Giampapa, M. & Mathews, M. (eds.): The Sun in Time, 118-158

STUIVER, M. (1994): Solar and climatic components of the atmospheric ^{14}C record. In: Frenzel, B.; Nanni, T.; Galli, M. & Gläser, B. (eds.): Solar output and climate during the Holocene. Paläoklimaforschung/Palaeoclimate Research 16 (this volume), 51-59

STUIVER, M. & BRAZIUNAS, T. F. (1993): Sun, ocean, climate and atmospheric $^{14}CO_2$, an evaluation of causal and spectral relationships. The Holocene 3, 289-305

WILLSON, R. C. & HUDSON, H. (1991): The sun's luminosity over a complete solar cycle. Nature 351, 42-44

Author's address:

Dr. E. Nesme-Ribes, URA 326, Observatoire de Paris, 5 Place Janssen, F-92195 Meudon

Climate sensitivity, Maunder Minimum and the Little Ice Age

Robert Sadourny

Summary

This paper discusses some aspects of the climate response to an increase (or decrease) of solar luminosity. In doing so, we have supposed that the time scale of solar flux variations is long enough to induce a significant warming (or cooling) of the oceanic mixed layer; this was indeed the case for the Maunder Minimum in the seventeenth century, which lasted at least fifty years. Assuming that solar luminosity was 0.4% less than its present value in the time period 1650-1700 A.D., model simulations indicate regional coolings of magnitude similar to observed data. Therefore, the Maunder Minimum is a plausible candidate for explaining the occurrence of the Little Ice Age, even though the influence of other mechanisms like volcanic eruptions or slow oceanic transients cannot be excluded. An interesting aspect of the response of climate to solar luminosity variations is the close similarity with its response to variations of the atmospheric loading in greenhouse gases. Therefore, if the Maunder Minimum is indeed the main cause of the Little Ice Age, the mean climate change from 1700 A.D. to present would be a good analogy of greenhouse warming. Like CO_2 doubling, an increase of solar luminosity is an external, global radiative perturbation deprived of geographical structure: characteristic of a very general class of climate warmings. A number of prominent features of such warmings can be listed, most of them seem robust enough to be only quantitatively affected by details of the physical formulations of the models. Among them, we may mention a decrease of the pole-to-equator temperature gradient in the lower troposphere due to snow-albedo feedback at high latitudes; a decrease of the temperature lapse rate in the tropical belt and the increase of the pole-to-equator moisture gradient, both due to strong low-level tropical moistening; a weakening of the mid-latitude cyclonic activity and the associated poleward transient enthalpy flux; a strengthening of the poleward transient moisture flux in mid-latitudes associated with increased precipitations in winter; a weaker Hadley circulation extending further poleward in the winter hemisphere; an enhanced drying of mid-latitude continents in summer; enhanced Monsoon precipitations with stronger Walker cell energetics.

Résumé

Nous étudions ici quelques aspects de la réponse du climat terrestre à une variation de la luminosité du soleil. Ce faisant, nous supposons que le temps caractéristique de cette variation est suffisamment long pour permettre un échauffement ou un refroidissement significatif de la couche mélangée océanique, ce qui fut effectivement le cas au XVIIe siècle pour le Minimum de Maunder, qui dura au moins cinquante ans. Supposant que la luminosité solaire était, entre 1650 et 1700, 0.4% plus faible que l'actuelle, nous obtenons un refroidissement d'un ordre de grandeur comparable aux données. Le Minimum de Maunder semble donc être un candidat plausible pour expliquer le Petit Age Glaciaire, même si des éruptions volcaniques ou des transitoires lents de l'océan ont pu influencer également le climat de cette période. Un aspect intéressant de la réponse du climat à des variations de luminosité solaire est qu'elle est très proche de la réponse à des variations de la charge atmosphérique en gaz à effet de serre. Si le Minimum de Maunder est bien à l'origine du Petit Age Glaciaire, le changement climatique entre 1700 A.D. et aujourd'hui serait donc un analogue utile du réchauffement par effet de serre. De même qu'un doublement du CO_2, une accroissement de la luminosité solaire est une perturbation radiative externe et globale, sans signature géographique, caractéristique d'une classe très générale de réchauffements climatiques. Nous mentionnerons ici quelques uns des traits principaux de ces réchauffements, dont la plupart paraissent suffisamment robustes pour n'être affectés que de manière quantitative par les détails des formulations physiques des modèles: l'affaiblissement du gradient de température pôle-équateur dans la basse troposphère, dû à la rétroaction neige-albedo aux hautes latitudes; l'affaiblissement de l'activité cyclonique hivernale des latitudes moyennes et du flux du chaleur associé vers le pôle; l'affaiblissement du gradient vertical de température dans les tropiques et le renforcement du gradient d'humidité pôle-équateur, tous deux dus à la forte humidification des basses couches dans les tropiques; le renforcement du flux d'humidité vers le pôle et des précipitations hivernales qui l'accompagnent aux latitudes moyennes; l'affaiblissement de la circulation de Hadley et son extension plus marquée vers le pôle d'hiver; une plus grande sécheresse continentale aux latitudes moyennes pendantl'été; enfin, des précipitations de mousson plus intenses, associés à un accroissement du transport d'énergie par la circulation de Walker.

1. Introduction

The climate of the earth is primarily regulated by both the global amount and the seasonal-latitudinal distribution of the energy it receives from the sun. On time scales of 10,000 to 100,000 years, it responds strongly to variations of orbital parameters which control insolation change, giving rise, according to the now relatively well established Milankovitch theory (e.g., MILANKOVITCH, 1930; BERGER, 1988), to the glacial-interglacial cycles which dominate the variability of the earth's climate during the last million years. The influence of variations of solar luminosity is less well established. With the time scales of billions of years, it must have had a major impact: we know that solar luminosity has increased by about one third since the formation of our planetary system, and that this increase has strongly conditioned the evolution of the earth's climate system in connection with variations of greenhouse forcing (GILLILAND, 1989). On time scales of tens to hundreds of million years, the crossing of galactical dust clouds by the solar system may have caused the recurrence of glacial ages. On shorter time scales of decades to centuries, however, variations of solar luminosity are rather small and their influence on climate still controversial (for a concise review of this problem, see SADOURNY, 1994b, for instance).

The earth's climate system can be considered as a complex oscillator which responds significantly to external perturbations only when these are strong enough and last long enough to be able to warm or cool the ocean by an appreciable amount. This is hardly the case for the eleven-year solar cycle: in addition to its short time scale, its amplitude in terms of integrated solar energy output is of the order of 0.5 W m^{-2} only (WILLSON & HUDSON, 1988). The current estimate sensitivity to an external (global) radiative perturbation is of the order of 0.4 to 1.2 (W m^{-2})$^{-1}$. When giving this estimate, one has to consider climate change as a simple equilibrium shift, a time scale of several decades being necessary to achieve the new equilibrium due to thermal inertia of the oceans (e.g., HOUGHTON et al., 1990). Because 30% of the incoming solar radiation is reflected back to space by the earth's surface, clouds, and aerosols, and because the absorbed part must be redistributed over the whole spherical surface, a solar luminosity variation of 0.5 W m^{-2} corresponds to an effective globally average radiative perturbation of the climate system of less than 0.1 W m^{-2}. Therefore, due to the weakness of its amplitude and the brevity of its period, the eleven-year cycle is unlikely to produce significant variations of the global climate. It is indeed remarkable that, in spite of such arguments, regional climatic fluctuations in phase with this solar cycle have been reported by several authors (e.g., SCHUURMANS, 1978; LABITZKE & VAN LOON, 1988, 1989; BARNSTON & LIVEZEY, 1989; MITCHELL, 1990).

The eighty-year Gleissberg cycle which modulates the amplitude of the eleven-year sunspot cycle would be a more plausible candidate for regulating the earth's climate, at least due to its longer time scale. The amplitude of the Gleissberg cycle in terms of energy variations is not precisely known, but could again be of the order of half a W m^{-2}. Such a perturbation would produce a global temperature response of 0.05 to 0.1°C, according to our present estimates of climate sensitivity. Several authors (REID, 1987; FRIIS-CHRISTENSEN & LASSEN, 1991) have indeed found interesting correlations between the modulation of the sunspot cycle and globally or hemispherically averaged surface temperatures during the last century, but the existence of a physical link remains conjectural.

The most attractive correlation so far between solar luminosity and climate seems to be the simultaneous occurrence of the Maunder Minimum (MAUNDER, 1894), which lasted at least fifty years, from the mid-seventeenth century to the early eighteenth century and the colder-than-now era called the Little Ice Age (for an account of the possible connection between the two events, see for example EDDY, 1976, 1977). In fact, even in this case, the connection is difficult to prove with quantitative physical arguments. On the one hand, the notion of the Little Ice Age concept is not always clearly defined, in terms of both duration and regional distribution. On the other hand, reconstructing the solar output deficit during the Maunder Minimum is a difficult matter. Recently, NESME-RIBES et al. (1993) (see also NESME-RIBES, this volume), using a combination of surrogate data and simplified solar dynamics theory, have estimated this deficit to be of the order of 5 to 6 Wm^{-2}. Again following the above estimate of climate sensitivity, the expected global temperature response would be a cooling of about half a degree to a degree, an estimate which could fit well the sparse temperature change data reported by several authors (LAMB, 1977; VAN DEN DOOL et al., 1978, LEGRAND et al., 1990; MATTHEWS, 1976; PFISTER, 1992). In view of such estimates, it is interesting to analyse the response of a three-dimensional climate model to a perturbation of the solar constant of that magnitude and compare it with historical data. This has been done recently using LMD atmospheric general circulation model (e.g., SADOURNY & LAVAL, 1984) in aversion which contains an interactive cloud-radiation scheme (LE TREUT & LI, 1991) and is coupled with a simplified slab ocean. These experiments have been reported and partially analysed in NESME-RIBES et al. (1993).

2. Climate warming and moistening from Maunder Minimum to present

When taking into account the albedo and sphericity of the earth, the effective radiative perturbation attributed to the Maunder Minimum (NESME-RIBES, this volume) is found to be in the order of 1 Wm^{-2}. Therefore, the change of climate from Maunder Minimum to present is roughly equivalent to the derivative of climate with respect to (or climate sensitivity to) radiative forcing. The following analysis is made with this equivalence in mind; in particular, most results are presented as differences between present climate and Maunder Minimum climate, which corresponds to a climate change of the warming type.

The simulations of NESME-RIBES et al. (1993) and parallel simulations of anthropogenic greenhouse warming show that the climate response to changes in the solar constant is almost undistinguishable from its response to a CO_2 increase of similar radiative magnitude, at least when considering the tropospheric response. This is probably explained by the fact that both perturbations, in addition to being purely radiative and having similar magnitude, have no geographical signature at all. The structure of the response, both in the horizontal and in the vertical, is imposed by internal feedbacks. The zonal and annual average warmings for an increase of the solar constant by 1.66% and a doubling of CO_2, both corresponding to an effective radiative perturbation of 4 Wm^{-2} at the top of the tropopause, are shown in Fig. 1. Surface warming in the tropics induces strong moistening there due to increased sensitivity exhibited by the Clausius-Clapeyron relation at warm temperatures (a warming of 5K at 25°C induces a 30% increase in saturation water vapour); this in turn decreases the vertical saturation lapse rate, inducing stronger warming at upper tropospheric levels. On the other hand, the atmospheric moistening induced by warming increases the greenhouse effect of water vapour and reduces the upwelling infrared flux: this explains the simulated stratospheric cooling also observed in the case of solar luminosity increase. The annual average geographical distributions of surface air temperature and moisture changes as simulated from Maunder Minimum to present are shown in Fig. 2. The latitudinal dependencies of the two fields are strikingly different: maximum warming occurs at high latitudes (due to snow-albedo feedback), while maximum moistening is observed in the tropics (due to the temperature dependency in Clausius-Clapeyron's law). There are important consequences of these two contrasting features.

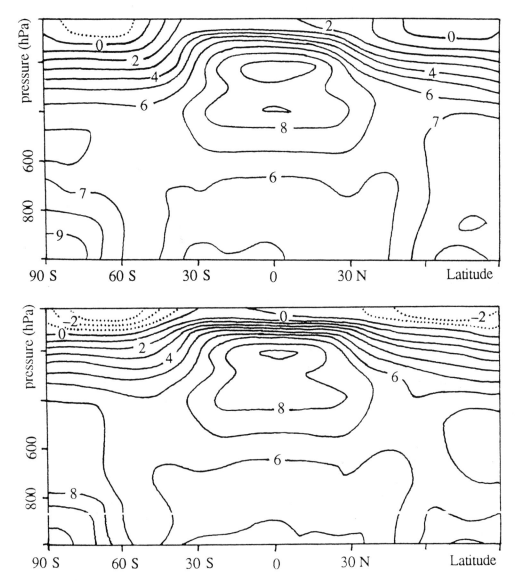

Fig. 1 Zonal and annual average meridional sections of temperature change for a 1.66% increase of solar irradiance (above) and a doubling of atmospheric CO_2 (below), in K

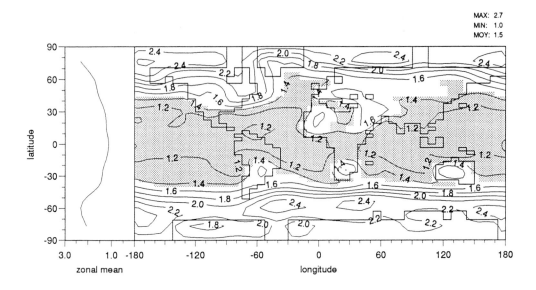

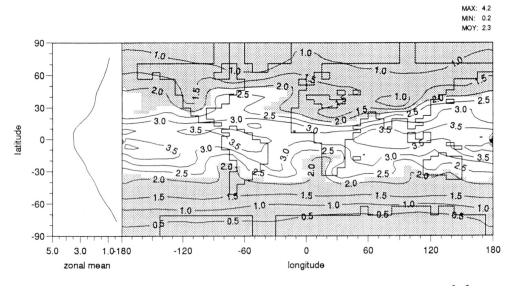

Fig. 2 Annual average distributions of surface air temperature change (above, in K W^{-2}m^2) and change of columnar moisture content (precipitable water, below, in mm W^{-1}m^2). Stippled areas indicate values less than global average. Zonal mean distributions on the left. (From Nesme-Ribes et al., 1993)

First, we may expect that baroclinic transient activity in midlatitude lower troposphere, associated with cyclogenesis and poleward enthalpy flux, has decreased from Maunder Minimum to present. Baroclinic transients in the lower troposphere are driven by the pole-to-equator temperature contrast at the surface: the magnitude of perturbation velocity is given by

$$v = \frac{g}{N} \frac{\Delta T}{T},$$ (1)

where g is the acceleration of gravity, N the Brunt-Vaisala frequency depending on vertical stability, T the mean surface temperature and ΔT the pole-to-equator temperature contrast. The mixing of entropy by baroclinic transients leads to the poleward enthalpy flux

$$v\Delta T = \frac{g}{N} \frac{(\Delta T)^2}{T}.$$ (2)

In the warming from Maunder Minimum to present, the decrease of the pole-to-equator temperature contrast yields a decrease of baroclinic activity:

$$\frac{\delta v}{v} = \frac{\delta \Delta T}{\Delta T} < 0$$ (3)

and an even stronger decrease of the poleward enthalpy flux:

$$\frac{\delta(v\Delta T)}{v\delta T} = 2\frac{\delta \Delta T}{\Delta T} < 0$$ (4)

On the other hand, the magnitude of the poleward moisture flux due to mixing of moisture by the baroclinic midlatitude transients, given by

$$v\Delta q = \frac{g}{N} \frac{\Delta T}{T} \Delta q,$$ (5)

where q refers to moisture, varies according to

$$\frac{\delta(v\Delta q)}{v\Delta q} = \frac{\delta \Delta T}{\Delta T} + \frac{\delta \Delta q}{\Delta q},$$ (6)

where the first term on the right-hand side is negative and the second positive. Comparing (4) with (6) yields that the poleward transport of energy under present conditions relies more heavily on water vapour (as compared to enthalpy) than it did under Maunder Minimum conditions. In fact, as shown by Fig. 3, the poleward moisture flux increases from Maunder Minimum to present: the effect of moisture contrast increase overcomes the effect of temperature contrast decrease in wintertime. As the accumulation of moisture in the atmosphere must be negligible, this means that precipitation in midlatitudes increases from Maunder Minimum to present.

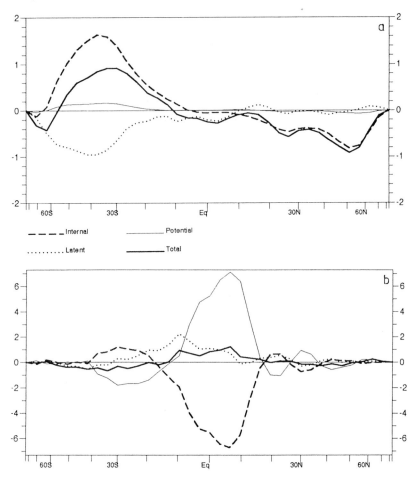

Fig. 3 Change of seasonal average meridional energy fluxes for June-July-August, due to transient eddies (a) and to mean meridional circulation (b). Bold continuous line: total energy; bold discontinuous line: enthalpy; dotted line: latent energy; thin continuous line: potential energy; units of $10^{14}m^2$. (From NESME-RIBES et al., 1993)

The picture is quite different in the upper troposphere. There the pole-to-equator temperature contrast increases, due to the strong moistening of the tropics and the resulting strong upper tropospheric warming already mentioned. This means stronger upper tropospheric westerlies in the subtropics, as shown clearly in Fig. 4. This strengthening of subtropical jets may have an interesting consequence on the water cycle. The poleward extent of the Hadley cell is limited by angular momentum dynamics (HELD & HOU, 1980): the stronger the upper tropospheric jets, the more poleward the Hadley cell reaches. One may conclude that the descending branch of the present Hadley cell is situated slightly more

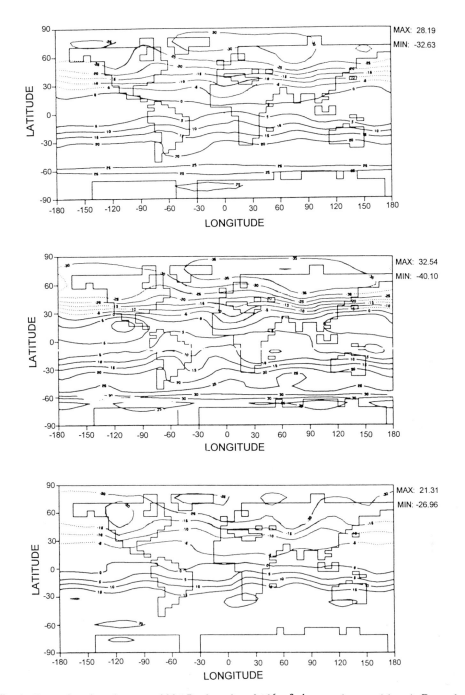

Fig. 4   Streamfunction change at 200 hPa, in units of 10^6 m^{2}s^{-1}; annual mean (above), December-January-February (middle); June-July-August (below). (From NESME-RIBES et al., 1993)

poleward than it was under Maunder Minimum conditions; this would mean a poleward shift of the desertic belt. However, these shifts are likely to be of small amplitude: the westerly increase over Sahara and North Africa, for example, is of the order of 5 m s^{-1}; this would cause a northward shift of about 50 km.

3. Response of the Hadley circulation and monsoons

The response of the Hadley cells in general is a matter of special interest, as it is intimately linked with tropical climate and monsoon precipitations. Hadley cells play an important dynamical role: in the tropics where angular momentum dynamics strongly homogenise the temperature field, horizontal mixing becomes inefficient and cannot transport energy from sources to sinks. The only way for the atmosphere to transport energy there is to organise itself into large-scale, quasi-stationary vertical rolls which transport energy in the direction of their upper branch. The mechanism of energy transport by a Hadley cell is as follows. The lower branch carries enthalpy and latent energy into the energy source region; moist convection transforms this enthalpy and this latent energy into potential energy by lifting the air parcels; potential energy is in turn carried away from the source region by the upper branch towards the subsiding branch in the energy sink region. Total energy is carried along the upper branch because, on average, the atmosphere is statically stable: the vertical gradient of total energy (the sum of enthalpy, latent energy and potential energy) is positive:

$$\frac{\delta}{\delta z}[c_pT + Lq + gz] > 0 \qquad (7)$$

Inequality (7), combined to the expression of energy flux

$$\mathbf{F} = \int_0^\infty (c_pT + Lq + gz)\,\rho\mathbf{U}dz\,, \qquad (8)$$

where ρ is density and $\mathbf{U}$ horizontal velocity, and to the equilibrium mass flux condition

$$\int_0^\infty \rho\mathbf{U}dz = 0\,, \qquad (9)$$

shows that the net, vertically integrated transport of energy in the Hadley (or Walker) cells occurs in the direction of the upper branch. In other words, because of vertical stratification, such large-scale overturnings are efficient ways for the atmosphere to transport energy in the absence of horizontal temperature gradients, from areas of ascending motion to areas of descending motion.

The response of the Hadley cell to radiative perturbations can thus be analysed from the viewpoint of energetic efficiency. The energetic efficiency of a Hadley cell can be defined as the ratio of the net energy exported to the sum of enthalpy and latent energy imported; in the present climate, it is of the order of 10%, and one may expect it to remain relatively stable while climate is changing. Indeed, during a warming, enthalpy and especially latent energy increase in the lower layers, but at the same time the tropopause is lifted and potential energy of the upper branch increases. Thus, if the efficiency remains stable, then the Hadley cell transports more energy for a given mass flux. As it is likely that the need to transport energy poleward does not increase (and even decreases, as the pole-to-equator temperature contrast decreases), this means that during climate warmings the mass flux of the Hadley cells must decrease: this is indeed verified in Fig. 5.

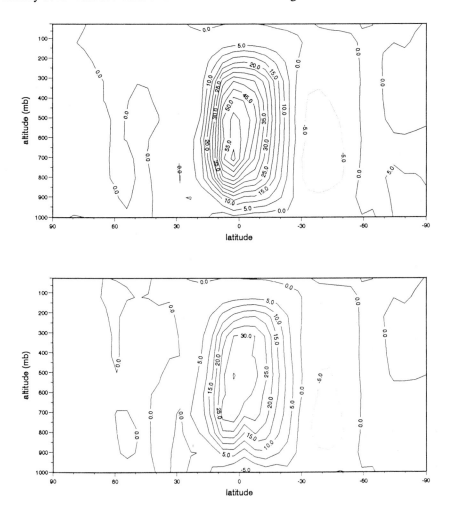

Fig. 5 June-July-August Hadley cells, visualized by a streamfunction in the latitude-pressure plane, for a climate warming experiment (absorbed solar flux increased by 4 W m^2: above) and a climate cooling experiment (absorbed solar flux decreased by 4 W m^2: below); unit: kPa m s^{-1}. (From SADOURNY, 1994a)

Another way to look at the response of the Hadley cell is to consider it as forced by the baroclinic term in the zonal component of the vorticity equation: the Hadley cell mass flux will increase if this source term, defined as

$$S = \frac{1}{\alpha} \left[\frac{\partial(\alpha,p)}{\partial(\varphi,z)} \right] = -\frac{R}{a} \left[\frac{\partial(T,Ln\,p)}{\partial(\varphi,z)} \right] \tag{10}$$

(where a is the earth's radius, φ is longitude, z altitude, α specific volume, p pressure, and the brackets refer to zonal averages), which maintains it against frictional dissipation, increases. In the tropics, where thermodynamic variables are well homogenised, we may replace zonally average products by products of averages without making appreciable errors. Expanding the latter expression in (10) and making use of the hydrostatic approximation, we get

$$S = \frac{R}{a} \left[\frac{\partial T}{\partial z} \right] \left[\frac{\partial Ln\,p}{\partial\varphi} \right] + \frac{g}{a} \left[\frac{\partial Ln\,T}{\partial\varphi} \right]. \tag{11}$$

If one considers the boreal Hadley cell, then both terms on the right-hand side are negative. In a warming process such as a transition from the Maunder Minimum conditions to present, we may not expect the meridional gradients of pressure and temperature to change significantly. The main variation is likely to come from the change in vertical lapse rate, triggered by lower layer moistening. The vertical lapse rate will decrease in absolute value, which means a decrease of the zonal vorticity source term, and therefore a weakening of the Hadley cell mass flux (see also SADOURNY, 1994a).

Monsoons can be considered a part of the Hadley-Walker circulation system, as they are generated by the need to export energy from overheated tropical continents in summer. Arguments similar to those above can thus be applied to analyse the response of monsoons to radiative perturbations such as a solar flux increase. Indeed, as the energy excess over the tropical continents in summer must increase with an increase of solar luminosity, we may expect an increase of the energy transport by the Monsoon systems. If, as noted above, the transport of energy by the Hadley cell decreases, this decrease must be more than compensated by an increase of the transport of energy by the Walker circulation. At the same time, we may expect an increase of Monsoon precipitations: first, because of the enhanced excess of continental energy in summer, and second, because of the strong tropical moistening of the lower layers which increases the ratio of latent heat flux to enthalpy flux in the lower layers of the Hadley-Walker circulations. This is indeed observed in NESME-RIBES et al. (1993) for the South-East Asia Monsoon.

4. Are simulations of the Maunder Minimum impact compatible with what we know of Little Ice Age conditions in Europe?

The climate of Europe during the Little Ice Age has been investigated by many authors who attempted to reconstruct temperature and precipitation using quantitative or proxy data at various locations (e.g., Central Europe, Switzerland, Norway, England and the Netherlands: LAMB, 1977; LEGRAND et al., 1990; MATTHEWS, 1976; PFISTER, 1992; SCHWEINGRUBER et al., 1978; VAN DEN DOOL et al., 1978). As noticed in Sadourny (1994a), the reconstructed cooling amounts (of the order of .5 to 2°C) are quite comparable to the simulations by NESME-RIBES et al. (1993) which are of the order of 1.5°C in the European area. The dispersion of simulated coolings in the model is, however, much smaller than the dispersion of reconstructed data, which may be explained by the rather coarse resolution used in the simulation.

These similarities between reconstructed and simulated coolings tend to indicate that the Maunder Minimum might be responsible for the Little Ice Age episode. However, we must be careful and avoid definite conclusions for a number of reasons: the model used in NESME-RIBES et al. (1993) is relatively sensitive to radiative perturbations (LE TREUT & LI, 1991), and the question of climate sensitivity and its dependency on cloud properties is far from solved (CESS et al., 1990); really convincing arguments on the involvement of the Maunder Minimum in the Little Ice Age conditions are still to come; finally, the causal relationship between Maunder Minimum and the Little Ice Age will be demonstrated only when other possible causes of climatic change, like an increase of volcanic activity (e.g., HANSEN & LACIS, 1990) or changes in the deep circulation of the ocean, are proved less relevant or less efficient in seventeenth century conditions.

5. Conclusion

The current estimate of solar irradiance change during the Maunder Minimum makes a climatic response of the order of current reconstructions of Little Ice Age cooling plausible. The Maunder Minimum thus appears as a strong candidate for explaining this cold episode in our recent climate, even though other perturbations like the thermohaline ocean circulation or volcanic eruptions may have helped in the process.

The main features of climate change produced by an increase of solar irradiance (as the one leading from Maunder Minimum to present conditions) can be described as follows: general warming, but stronger at high latitudes due to the snow-albedo feedback; general moistening, but stronger in the tropics, due to the Clausius-Clapeyron law; a weaker tropospheric temperature lapse rate in the tropics; enhanced midlatitude surface drying in summer, in direct response to the warming; weakening of midlatitude transient cyclonic activity in winter, associated with a decrease of the wintertime poleward energy flux, both due to the decrease of the pole-to-equator temperature contrast; increase of the poleward moisture flux and associated midlatitude precipitations in winter, due to the increase of the pole-to-equator moisture contrast; weaker mass fluxes and further poleward extension of the winter hemisphere Hadley cell; and finally, an increase of Walker cell energetics and Monsoon precipitation. Currently available data on the Little Ice Age, however, are much too scarce to validate most of these predicted changes.

References

BARNSTON, A. G. & LIVEZEY, R. E. (1989): A closer look at the effect of the 11-year solar cycle and the quasi-biennial oscillation on northern hemisphere 700 mb height and extratropical North American surface temperatures. J. Climate 2, 1295-1313

BERGER, A. (1988): Milankovitch theory and climate. Review of Geophysics 26, 624-657

CESS, R. D.; POTTER, G. L.; BLANCHET, J. P.; BOER, G. J.; DEL GENIO, A. D.; DEQUE, M.; DYMNIKOV, V.; GALIN, V.; GATES, W. L.; GHAN, S. J.; KIEHL, J. T.; LACIS, A. A.; LE TREUT, H.; LI, Z. X.; LIANG, B. J.; MCAVANEY, ; MELESHKO, V. P.; MITCHELL, J. F. B.; MORCRETTE, J. J.; RANDALL, D. A.; RIKUS, L.; ROECKNER, E.; ROYER, J. F.; SCHLESE, U.; SHEININ, D. A.; SLINGO, A.; SOKOLOV, A. P.; TAYLOR, E. E.; WASHINGTON, W. M.; WETHERALD, R. T.; YAGAI, I. & ZHANG, M. H. (1990): Intercomparison and interpretation of climate feedback processes in nineteen atmospheric general circulation models. J. Geophys. Res. 95, 16601-16615

EDDY, J. A. (1976): The Maunder Minimum. Science 192, 1189-1190

EDDY, J. A. (1977): Climate and the changing sun. Climatic Change 1, 173-190

FRIIS-CHRISTENSEN, E. & LASSEN, K. (1991): Length of the solar cycle: an indicator of solar activity closely associated with climate. Science 254, 698-700

GILLILAND, R. L. (1989): Solar evolution. Global and Planetary Change 1, 35-55

HANSEN, J. & LACIS, A. A. (1990): Sun and dust *versus* greenhouse gases: an assessment of their relative roles in global climate change. Nature 346, 713-719

HELD, I. M. & HOU, A. Y. (1980): Non-linear axially symmetric circulations in a nearly inviscid atmosphere. J. Atmosph. Sci. 37, 515-533

HOUGHTON, J. T.; JENKINS, G. J. & EPHRAUMS, J. J. (eds.) (1990): Climate Change. The IPCC Scientific Assessment. Cambridge University Press

LABITZKE, K. B. & VAN LOON, H. (1988): Association between the 11-year solar cycle, the QBO, and the atmosphere, Part I: The troposphere and stratosphere of the Northern Hemisphere in winter. J. Atm. Terr. Phys. 50, 197-206

LABITZKE, K. B. & VAN LOON, H. (1989):Association between the 11-year solar cycle, the QBO, and the atmosphere, Part III: Aspects of the association. J. Climate 2, 554-565

LAMB, H. H. (1977): Climate: present, past and future, Vol. 2: Climatic history and the future. Methuen, London, 835 p.

LEGRAND, J.-P.; LE GOFF, M. & MAZAUDIER, C. (1990): On the climatic changes and the sunspot activity during the seventeenth century. Annales Geophysicae 8, 637-644

LE TREUT, H. & LI, Z. X. (1991): Sensitivity of an atmospheric general circulation model to prescribed SST changes: feedback effects associated with the simulation of cloud optical properties. Climate Dynamics 5, 175-187

LI, Z. X. & LE TREUT, H. (1992): Cloud-radiation feedback in a general circulation model and its dependence on cloud modelling assumptions. Climate Dynamics 7, 133-139

MATTHEWS, J. A. (1976): "Little Ice Age" palaeotemperatures from high altitude tree growth in south Norway. Nature 264, 243-245

MAUNDER, T. (1894): A prolonged sunspot minimum. Knowledge 17, 173-176

MILANKOVITCH, M. (1930): Mathematische Klimalehre und Astronomische Theorie der Klimaschwankungen, Gebrüder Bornträger, Berlin

MITCHELL, J. M. Jr. (1990): Climate variability: past, present and future. Climatic Change 16, 231-246

NESME-RIBES, E.; FERREIRA, E. N.; SADOURNY, R.; LE TREUT, H. & LI, Z. X. (1993): Solar dynamics and its impact on solar irradiance and the terrestrial climate. J. Geophys. Res. 98, 923-935

NESME-RIBES, E. (1995): The Maunder Minimum and the deepest phase of the Little Ice Age: a causal relationship or a coincidence? In: Frenzel, B.; Nanni, T.; Galli, M. & Gläser, B. (eds.): Solar output and climate during the Holocene. Paläoklimaforschung/ Palaeoclimate Research 16 (this volume), 131-144

PFISTER, C. (1992): Monthly temperature and precipitation in central Europe 1525-1979: quantifying documentary evidence on weather and its effects. In: Bradley, R. S. & Jones, P. D. (eds.): Climate since A.D. 1500. Routledge, London, 118-141

REID, G. C. (1987): Influence of solar activity on global sea surface temperature. Nature 329, 142-143

SADOURNY, R. (1994a): Maunder Minimum and the Little Ice Age: impact of a long-term variation of the solar flux on the energy and water cycle. In: Duplessy, J.-C. & Spyridiakis, M. T. (eds.): Long-term climatic variations. NATO-ASI Series I-22, Springer Verlag, Berlin, Heidelberg

SADOURNY, R. (1994b): L'influence du Soleil sur le climat. C. R. Acad. Sci. Paris 319-II, 1325-1342

SADOURNY, R. & LAVAL, K. (1984): January and July performance of the LMD general circulation model. In: Berger, A. & Nicolis, C. (eds.): New perspectives in climate modelling, Elsevier

SCHUURMANS, C. J. E. (1978): Influence of solar activity on winter temperatures: new climatological evidences. Climatic Change 1, 231-238

SCHWEINGRUBER, F. H.; FRITTS, H. C.; BRÄKER, O. U.; DREW, L. G. & SCHÄR, E. (1978): The X-ray technique as applied to dendrochronology. Tree-Ring Bulletin 38, 61-91

VAN DEN DOOL, H. M.; KRIJNEN, H. J. & SCHUURMANS, C. J. E. (1978): Average winter temperatures at de Bilt (the Netherlands): 1634-1977. Climatic Change 1, 319-330

WILLSON, R. C. & HUDSON (1988): Solar luminosity variations in solar cycle 21. Nature 332, 810-812

Author's address:

Dr. R. Sadourny, Laboratoire de Méteorologie Dynamique, Ecole Normale Supérieure, 24 rue Lhomond, F-75231 Paris Cedex 05

Solar radiation and global climate change: some experiments with a two-dimensional climate model

Thierry Fichefet

Summary

Numerical experiments are carried out with a two-dimensional climate model in order to elucidate the possible relative importance of the solar and astronomical forcings on the climate of the last three centuries. The model domain covers the northern hemisphere only. The atmosphere is represented by a zonally averaged quasi-geostrophic model including a new formulation of the meridional transport of quasi-geostrophic potential vorticity and a parameterization of the Hadley sensible heat transport. At the surface, the model has land-sea resolution and incorporates detailed snow and sea-ice mass budgets. The upper ocean is represented by an integral mixed-layer model in which meridional convergence of heat is parameterized by a diffusive law. A comparison between the computed and observed present-day climates indicates that the model does reasonably well in simulating the seasonal cycle of various climatic variables. In a study of sensitivity to changes in the solar constant, the model exhibits a relatively high degree of non-linearity. The change in the equilibrium surface temperature is +3.2°C in response to a 2% increase in the solar constant and -3.7°C in response to 2% decrease in the solar constant. When the total solar irradiance is reduced by 0.25% (which is a plausible estimate of the change in the sun's radiative output during the Maunder Minimum), the equilibrium surface temperature decreases by 0.4°C. The model is then coupled to a diffusive deep-ocean model in order to investigate the transient response of climate to the solar and astronomical forcings between the pre-industrial era and the present time. The solar irradiance changes are parameterized from recent satellite observations using the Wolf number as a basis, whereas the variations in solar radiation due to the changes in the earth's orbital elements are taken from BERGER (1978). In the model the former forcing induces a warming of about 0.005°C between the time intervals 1765-1875 and 1876-1990, while the latter is responsible for a cooling of about 0.003°C. These changes appear weak compared to the greenhouse-gas-induced warming simulated by the model.

Résumé

Des expériences numériques sont réalisées à l'aide d'un modèle bidimensionnel du système climatique dans le but d'évaluer l'influence des forçages solaire et astronomique sur le climat des trois derniers siècles. Le modèle couvre uniquement l'hémisphère nord. L'atmosphère est représentée par un modèle quasi-géostrophique moyenné zonalement qui inclut une nouvelle formulation du transport méridien de vorticité potentielle quasi-géostrophique ainsi qu'une paramétrisation du transport de chaleur sensible par la cellule de Hadley. En surface, le modèle possède une résolution terre-mer et calcule explicitement les bilans massiques de neige et de glace. L'océan superficiel est représenté par un modèle intégral de la couche mélangée dans lequel la convergence méridienne de chaleur est paramétrisée par une loi diffusive. Une analyse du climat simulé par le modèle pour les conditions présentes indique qu'il reproduit plus ou moins correctement les grandes caractéristiques du cycle saisonnier du climat de l'hémisphère nord. La sensibilité du modèle à un changement de la constante solaire est alors investiguée. Cette étude montre que le modèle présente un comportement non-linéaire. A l'équilibre, le changement de température de surface prédit pour une augmentation de la constante solaire de 2% s'élève à +3.2°C, alors qu'il est de -3.7°C pour une diminution équivalente de la constante solaire. Lorsque la constante solaire est réduite de 0.25% (ce qui est une estimation plausible du changement d'intensité solaire durant le Minimum de Maunder), la température de surface d'équilibre diminue de 0.4°C. Le modèle est ensuite couplé à un modèle diffusif de l'océan profond pour étudier la réponse transitoire du climat aux forçages solaire et astronomique de l'ère pré-industrielle à nos jours. Les changements d'intensité solaire sont obtenus à partir d'une série temporelle du nombre de Wolf en utilisant une paramétrisation basée sur des observations satellitaires récentes, tandis que les variations de l'éclairement liées aux modifications des éléments orbitaux de la Terre sont déduites de BERGER (1978). Le premier forçage induit dans notre modèle un réchauffement d'environ 0.005°C entre les périodes 1765-1875 et 1876-1990, alors que le second est responsable d'un refroidissement de 0.003°C. Ces changements paraissent faibles par rapport au réchauffement simulé en réponse à l'augmentation de la concentration des gaz à effet de serre.

1. Introduction

Global climate change is becoming one of the major scientific issues of the decade, largely because it has a potentially greater impact on the world's society than almost any other phenomenon and because civilization itself may well be a root cause through anthropogenic changes in the composition of the atmosphere. Unfortunately, the climatic effects of these changes have to be detected against the background noise of natural climate variability, whose nature and causes are still uncertain and controversial.

Since the sun is the principal energy source of the earth's climatic system, it is natural to suspect variations in solar radiation as a possible source of climate variations at the secular time scale (REID, 1991). There are two distinct causes of this variability of solar radiation (HOUGHTON et al., 1990). The first is related to the periodic changes in the earth's orbital elements, which in turn influence the latitudinal and seasonal distribution of the solar energy received by the earth (the so-called "astronomical or Milankovitch effect"). These changes act with greatest impact on time scales of 10,000 to 100,000 years and have been responsible, at least partly, for the major glacial-interglacial cycles that occurred during the

Quaternary period (BERGER, 1988). The second source of variability of solar radiation comes from physical changes of the sun itself; such changes occur on almost all time scales. Models of stellar evolution indicate that the solar luminosity has increased uniformly over geologic time. As hydrogen is converted to helium in the solar core, the rate of nuclear burning increases, causing the total solar irradiance (also called the "solar constant") to be some 20 to 30% higher today than it was at the time of the earth's formation (NEWMAN & ROOD, 1977). Continuous satellite measurements of total solar irradiance have been made since 1978. These have shown that, on time scales of a few days to one decade, there are irradiance variations that are associated with activities in the sun's outer layer, the photosphere - specifically, sunspots and bright areas known as faculae (LEAN, 1991). The very high frequency changes are too rapid to affect the climate noticeably. However, there is a lower frequency component that follows the 11-year sunspot cycle which may have a climatic effect. Indeed, it has been found that the total solar irradiance during the sunspot minimum of 1986 was about 0.1% less than during the previous (1980) maximum (WILLSON & HUDSON, 1988) and that the changes since 1986 have continued to parallel the 11-year sunspot cycle (WILLSON & HUDSON, 1991). These recent findings have led to the development of a series of empirical models relating total irradiance to photospheric magnetic activity (e.g., SCHATTEN, 1988; WILLSON & HUDSON, 1988; FOUKAL & LEAN, 1990).

In the present study, numerical experiments are conducted with a two-dimensional zonally averaged climate model in order to quantify the possible effects of the astronomical forcing and of the solar irradiance changes derived from the empirical model of WILLSON & HUDSON (1988) on the climate of the last three centuries. Experiments are also performed to compare these effects to those induced by the observed forcing due to the gradual increase in greenhouse-gas concentrations. It is worth noting that the model used here exhibits no internal variability, which allows an unequivocal quantitative assessment of the solar and astronomical effects. Of course, this model's peculiarity has to be borne in mind when extending the findings of the present study to the real world.

The paper is organized as follows. The climate model is described in Section 2. Section 3 compares selected results of a simulation of the present climate with observations. In Section 4, we examine the model sensitivity to changes in the solar constant. Section 5 deals with the transient response of the model to the solar, astronomical, and greenhouse-gas forcings between the pre-industrial era and the present time. Concluding remarks are finally given in Section 6.

2. Model description

Our model is a two-dimensional zonally averaged climate model specifically designed for simulating the seasonal cycle of the climate of the northern hemisphere. A summary of the model characteristics is given here; further details can be found in GALLÉE et al. (1991).

The atmospheric component of the model is based on the zonally averaged form of the two-level quasi-geostrophic potential vorticity system of equations (including diabatic heating and frictional dissipation) written in spherical and pressure coordinates (SELA & WIIN-NIELSEN, 1971; OHRING & ADLER, 1978). The basic output consists of the latitudinal distributions of the temperature at 500 hPa and of the zonal winds at 250 and 750 hPa. Meridional transport of quasi-geostrophic potential vorticity is accomplished by an eddy mixing process using exchange coefficients parameterized as in GALLÉE et al. (1991). It is well known that the quasi-geostrophic approximation leads to an underestimation of the strength of the Hadley cell in low latitudes and, therefore, to an overestimation of the temperatures in these regions (e.g., WHITE & GREEN, 1984). To partly remedy this problem, a parameterization of the vertically integrated Hadley heat transport (PENG et al., 1987) has been introduced in the model. Separate surface energy balances are calculated over various surface types at each latitude (see below), and the heating of the atmosphere due to the vertical heat fluxes is the weighted average of the convergences of these fluxes above each kind of surface. The vertical heating processes considered are: solar radiation, long-wave radiation, convection, and latent heat release.

The solar radiation scheme used here is very similar to the one described by TRICOT & BERGER (1988). The following processes are included: absorption by H_2O, CO_2, and O_3, Rayleigh scattering, absorption and scattering by cloud droplets and aerosols, and reflection by the earth's surface. The long-wave radiation computations are based on MORCRETTE's (1984) wide-band formulation of the radiative transfer equation. Absorption by H_2O, CO_2, and O_3 is explicitly treated and the cloud cover is supposed to behave as a blackbody, as is the earth's surface. In both schemes, account is taken of variation in surface topography.

A single cloud layer is assumed to exist in each latitude belt, with monthly cloud amount prescribed from zonal mean climatology. The base and top altitudes of the cloud layer and its optical thickness are kept fixed throughout the year. The surface fluxes of sensible and latent heat are parameterized according to SALTZMAN & ASHE (1976) and SALTZMAN (1980), respectively. Latent heat release in the atmosphere is obtained from observed zonal and monthly mean precipitation rates uniformly scaled to ensure a balance between precipitation and evaporation over the whole hemisphere. Above the Greenland ice sheet, precipitation rates are further modified to incorporate the effects induced by the distance to the moisture source and the surface slope and elevation of the ice sheet (OERLEMANS, 1982).

At each latitude, the model surface is resolved into continental and oceanic portions. Land can be partly covered by snow and the Greenland ice sheet. The mean altitude of the ice-free land and the elevation and extent of the Greenland ice sheet are specified from data. Sea ice can exist at the ocean surface and vary in amount at the expense of the open-water area.

Over the land free of snow and ice, the surface temperature is determined by an energy-balance equation accounting for the subsurface heat storage (TAYLOR, 1976). The groundwater budget is computed by the so-called "bucket method" of MANABE (1969). The soil is assumed to have a water-holding capacity of 40 cm. If the calculated soil moisture exceeds this value, the surplus is supposed to run off. Changes in soil moisture depend on the rates of rainfall, evaporation, snow melt, and run-off. The fraction of the ice-free land area above which precipitation falls as snow, f_{sf}, is parameterized as a function of the surface temperature following HARVEY (1988a). As the surface temperature decreases, f_{sf} increases but only a single average snow depth is stored. When f_{sf} decreases from one time step to the next, the new snow is uniformly distributed over the existing snow-covered area. The surface temperature of the snow layer is obtained by assuming an equilibrium between the internal conductive heat flux and the other surface heat fluxes. When the predicted temperature reaches 0°C, snow melt occurs. This melting is partitioned between a decrease in the snow extent and a decrease in the snow depth, so that the snow area is gradually reduced during the melting period. The Greenland ice sheet is supposed to be completely covered by perennial snow. The proportion of precipitation falling as snow over its surface is a function of the surface temperature as in LEDLEY (1985). An energy-balance equation similar to the one applied for snow on land provides the temperature at the air-snow interface. The net accumulation rate of snow is derived from an explicit surface mass budget.

The upper ocean (up to a depth of 150 m) is represented by the variable depth and temperature mixed-layer model of GASPAR (1988). Meridional convergence of heat due to oceanic currents is modelled by a diffusive law according to SELLERS (1973). In the presence of sea ice, the mixed-layer depth is fixed at 30 m. The surface temperature of the ice pack and the vertical ice growth and decay rates are computed by the zero-layer model of SEMTNER (1976). Snow on top of the ice layer is not taken into account and no dynamic effect, such as ice drifting caused by wind, is considered. The opening and closing of leads and the open-water temperature are calculated following the technique of PARKINSON & WASHINGTON (1979), subject to small modifications described in GALLÉE et al. (1991). The heat flux from the ocean to the ice is computed by assuming that sea ice is in thermodynamic equilibrium with the water just below. Therefore, both the temperature at the base of the ice and the temperature of the under-ice water are supposed to be equal to the freezing point of sea water. In the model, the only physical mechanism that can disturb this equilibrium is a partial mixing between the under-ice water and the lead water. To maintain the ice-covered water at the freezing point, the heat flux from the ocean to the ice has to compensate exactly for that perturbation.

Separate albedos are calculated for each surface type. The albedo of the land not covered by snow or ice is a function of the soil moisture content according to SALTZMAN & ASHE (1976). Over continental areas covered by forests, the snow albedo is set equal to 0.40 (ROBOCK, 1980). Elsewhere, the snow albedo is dependent on the surface temperature and the snow age as in DANARD et al. (1984). Values of 0.85 and 0.65 are assigned to the maximum and minimum albedos of dry snow, respectively. For melting snow, the upper and lower limits of the albedo are assumed to be 0.70 and 0.50, respectively. These albedos are corrected later on to include the effects of snow depth and solar zenith angle. The albedo of the ice-free ocean under clear sky is parameterized as a function of the solar zenith angle after BRIEGLEB & RAMANATHAN (1982). Under cloudy conditions, the sea surface has an albedo of 0.07. The albedo of sea ice is supposed to vary linearly with surface temperature, from 0.75 at -10°C to 0.40 at 0°C. A solar zenith angle correction is also applied here (ROBOCK, 1980). In all the experiments discussed below, a latitudinal resolution of 5° and a time interval of 3 days were used.

3. Simulation of the present climate

The model described above was first employed to simulate the present climate of the northern hemisphere. This control experiment was performed with a CO_2 concentration of 330 parts per million by volume (ppmv). Referring to WILLSON et al. (1981), the solar constant, S_o, was taken as 1368 Wm^{-2}. Starting from arbitrary initial conditions, the model was run until an equilibrium seasonal cycle was established. The equilibrium was supposed to be achieved when the hemispheric annual mean radiative balance at the top of the atmosphere became less than 0.01 Wm^{-2}. This was reached after 50 years of integration.

The seasonal cycle of the 500 hPa temperature simulated by the model is depicted in Fig. 1a. The magnitude and horizontal gradient of temperature are in general agreement with the monthly mean data of OORT (1983). A plot of the difference between the estimated and observed temperatures in Fig. 1b reveals that the major discrepancy occurs during winter in the tropics, where the modelled temperature is underestimated by 4°C. Reasons for this error basically stem from the limitations of the Hadley heat transport formulation. In the middle and high latitudes, the computed temperature is 3 to 4°C too low during spring and early summer. This deficiency is associated with an overestimation of the snow extent at the beginning of the melting period (see below).

Figure 2a illustrates the seasonal variation of the oceanic mixed-layer temperature predicted by the model. The departure of the modelled temperature from the observed one (LEVITUS, 1982) is displayed in Fig. 2b. This figure shows that the temperature difference never exceeds 2°C in the tropics. In the vicinity of 50°N, the computed temperature is 4°C too high during summer. This feature is mainly attributed to an insufficient vertical mixing of the spring and summer surface heat input, which is caused by an underestimation of the surface wind speed. Around 70°N, the model temperature is lower than observed by 4°C from May to October. An examination of the seasonal cycle of the sea-ice concentration simulated by the model indicates that the sea-ice compactness is overestimated at this latitude in spring. The overextensive ice cover prevents solar radiation from penetrating into the water, and thus delays and weakens the summer warming of the ocean. This effect, together with an exaggerated divergence of the oceanic heat transport in autumn, explains most of the temperature discrepancy. North of 80°N, the mixed layer remains close to the freezing point throughout the year, in agreement with the data.

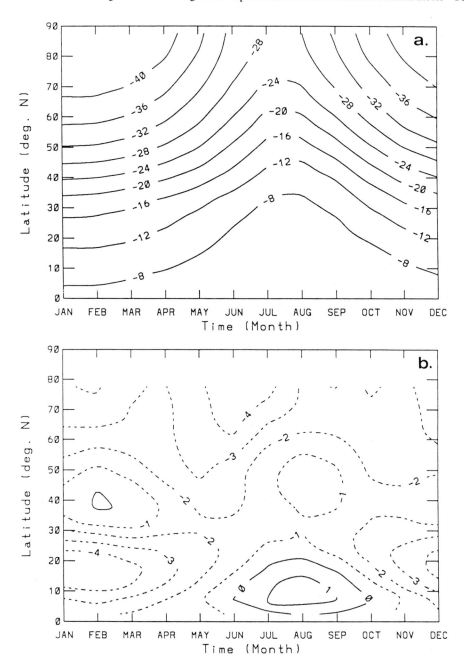

Fig. 1 Latitude-time distribution of the zonal mean temperature at 500 hPa simulated by the model (a) and of the difference between the computed temperature and the observations of OORT (1983) (b). Units are °C. Dashed lines denote negative deviations

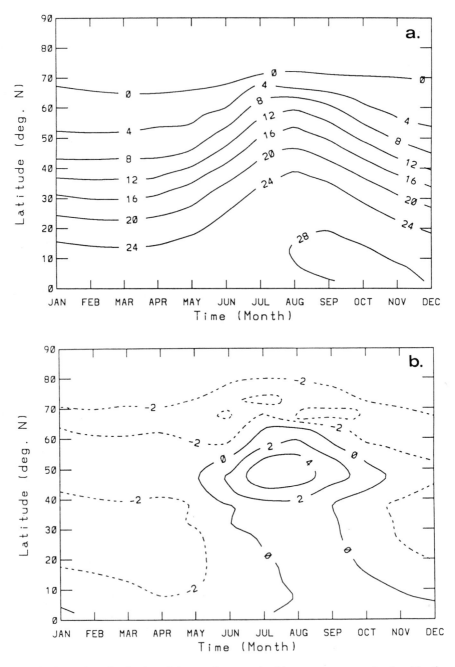

Fig. 2 Latitude-time distribution of the zonal mean mixed-layer temperature simulated by the model (a) and of the difference between the computed temperature and the observations of Levitus (1982) (b). Units are °C. Dashed lines denote negative deviations

Due to its high albedo, the cryosphere plays a crucial role in the energy balance of the earth's climate system. Therefore, a necessary condition for a reliable numerical prediction of climate change is that the sea-ice and snow simulations in the control run are realistic (WILSON & MITCHELL, 1987). Figure 3a compares the seasonal cycle of the total sea-ice area simulated by the model with the observations of ROBOCK (1980). It can be seen that the model reproduces the amplitude of the observed cycle fairly well. However, the ice growth is significantly underestimated in early autumn and the maximum ice extent occurs one month too late. As a consequence, there tends to be too little ice from October to March and slightly too much from May to August. Figure 3b demonstrates that the model does reasonably well in simulating the seasonal cycle of the continental snow area. Some small discrepancies between the model and data are, however, visible. In particular, the computed snow extent does not increase fast enough in early autumn. As a result, the maximum snow area predicted by the model appears a bit overestimated compared to the data. Note also that the model slightly underestimates the melting rate of snow in early spring.

Further evidence of the model performance for present-day climate is given in GALLÉE et al. (1991). That paper demonstrates that the model shows acceptably good agreement with enough aspects of the seasonal behaviour of the real climate system to permit useful studies of a range of possible environmental perturbations.

4. Model sensitivity to changes in the solar constant

In order to test the model sensitivity to changes in the sun's radiative output, we have evaluated its equilibrium response to both a 2% increase and a 2% decrease in the solar constant. Convergence to equilibrium for these experiments was achieved after integrating the model over a period of 50 years, starting from initial conditions similar to the ones used in the control run discussed in the previous section. Table 1 summarizes the results obtained.

In response to a 2% increase in the solar constant, the hemispheric annual mean surface temperature simulated by the model increases by 3.2°C. It is worth pointing out that this temperature sensitivity lies within the range of current estimates (e.g., WETHERALD & MANABE, 1975; PENG et al., 1982; HANSEN et al., 1984; POTTER & CESS, 1984; HARVEY, 1988b; NEEMAN et al., 1988). Table 1 shows that the magnitude of the annual mean surface warming increases with latitude, resulting in a general reduction of the meridional temperature gradient. This feature is due primarily to the sea-ice and snow albedo/temperature feedback. This feedback mechanism requires sufficient warming to cause substantial melting of ice and snow in middle and high latitudes. The resulting decrease of albedo then allows further warming, more melting of ice and snow, and so on. Note that there is a reduction in the maximum sea-ice area of about 3×10^6 km^2 in the +2% S$_o$ case, and that the ice pack is completely removed from the Arctic Ocean during summer in this experiment.

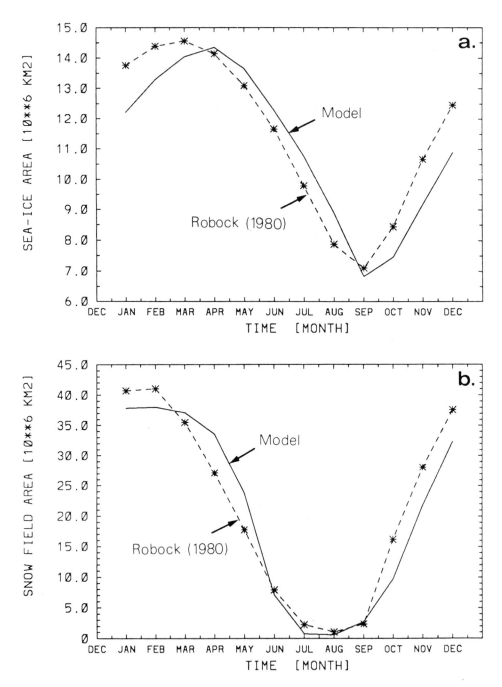

Fig. 3 Comparison of modelled and observed seasonal cycle of (a) total sea-ice area and (b) total continental snow area. The observations are from Robock (1980)

As far as the continental snow is concerned, the maximum snow extent decreases by about 6×10^6 km^2. When the solar constant is reduced by 2%, the hemispheric annual mean surface temperature predicted by the model decreases by 3.7°C. This larger sensitivity, compared to the previous experiment, is a consequence of the non-linear character of the earth's climate system (e.g., POTTER & CESS, 1984; NEEMAN et al., 1988). Table 1 indicates that, once again, the temperature response is enhanced at high latitudes.

Table 1 Model sensitivity to changes in the solar constant S_o. For the +2% S_o and -2% S_o cases, the table gives the surface temperature response relative to the control run. N.H. refers to as Northern Hemisphere

	Annual Mean Surface Temperature (°C)					Sea-Ice Area (10^6 km^2)		Snow Area (10^6 km^2)	
	N.H.	7.5°N	52.5°N	62.5°N	67.5°N	max	min	max	min
Control Run	15.3	27.2	4.0	-4.5	-9.8	14.4	6.8	38.0	0.60
+2% S_o	+3.2	+2.9	+3.2	+3.6	+4.0	11.6	0.0	32.1	0.10
-2% S_o	-3.7	-3.1	-3.7	-4.9	-4.9	18.8	11.4	44.3	2.34

Using Ca II H- and K-line emission as an index of the sun's brightness, LEAN et al. (1992) matched and analysed the emissions observed for non-cycling stars, finding that the total solar irradiance may have decreased below its current mean level by about 0.25% during the Maunder Minimum of sunspot activity (1645-1715). A perturbation of this magnitude generates in our model a hemispheric annual mean surface cooling of 0.4°C. Comparing this with the temperature decrease of approximately 1°C that occurred during the Little Ice Age suggests that the solar forcing may have contributed some, but not all, of the observed warming from the Little Ice Age to the present time. Figure 4 gives the latitude-time distribution of the zonally averaged surface temperature response. In the tropics the magnitude of the cooling is less than 0.4°C and depends little upon season. At mid-latitudes, 0.4-0.5°C coolings take place during spring and early summer. These relatively large coolings are due to a delay in the melting of the snow cover over land and to the resulting albedo/temperature feedback. In the polar latitudes, the surface temperature response exhibits a strong seasonal variation. The cooling is intense (up to 1.5°C) in autumn and early winter, whereas it is weak (less than 0.2°C) during summer months. This weak summer response can be explained by the fact that summer sea ice is at the melting point in both the Maunder Minimum experiment and in the control run. On the other hand, as the solar constant decreases, the rate of ice melting decreases in summer. This causes the ice pack to be thicker at the start of winter, which leads to an enhanced insolation effect responsible for the large cooling simulated in autumn and early winter.

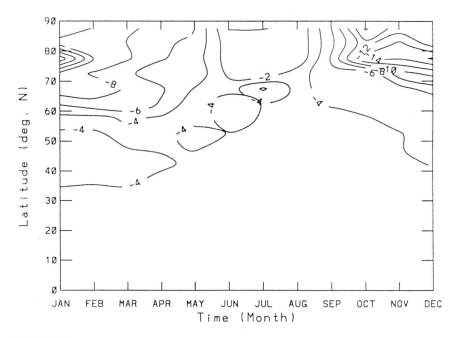

Fig. 4 Equilibrium response of the zonally averaged surface temperature to a 0.25% decrease in the solar constant. The response is relative to the control run. Units are tenths of °C

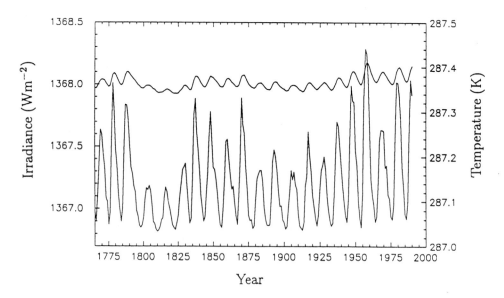

Fig. 5 Reconstructed solar irradiance changes (lower curve; see Section 5.1 for explanations) and time evolution of the hemispheric annual mean surface temperature simulated by the model in response to this forcing (upper curve)

5. Simulation of the transient response of climate to the solar, astronomical, and greenhouse-gas forcings over the last three centuries

In this section, we examine the model transient response to reconstructed time-series of solar radiation between the pre-industrial era and the present time. We consider the solar irradiance changes due to the observed variations in the photospheric magnetic activity, as well as the variations in solar radiation caused by the changes in the earth's orbital elements. We also investigate the model response to the build-up of atmospheric greenhouse gases resulting from human activities.

In all the experiments presented here, an equilibrium climate condition was first established for the total solar irradiance, the earth's orbital elements, and the atmospheric composition corresponding to the year 1700. Using this equilibrium as the initial condition, the model was then applied up to 1990, with the forcings changing with time according to plausible reconstructions. During this stage of the experiments, the uptake of heat perturbations by the deep ocean was approximated as a vertical diffusion process (HANSEN et al., 1988; TRICOT, 1989). The deep ocean, taken to be the water below the maximum allowable mixed-layer depth (i.e., 150 m), is divided into 13 layers of geometrically increasing thickness, with a total thickness of 3850 m. The latitude-dependent effective diffusion coefficient, k, is deduced from HANSEN et al. (1984) and ranges from 0.2 cm^{2}s^{-1} near the equator to 5 cm^{2}s^{-1} in high latitudes, leading to a hemispheric mean value of about 1.5 cm^{2}s^{-1}. Note that k is constant in time and in the vertical direction. It is obvious that this simple treatment of the deep ocean excludes the effects of natural variability associated with ocean dynamics (e.g., MIKOLAJEWICZ & MAIER-REIMER, 1990; MANABE et al., 1991; MYSAK et al., 1993) and the possibility of switches between the basic modes of ocean circulation (e.g., BROECKER, 1991; MAROTZKE & WILLEBRAND, 1991; STOCKER & WRIGHT, 1991). Thus, it must be considered as being only a first step in studying the transient response of climate to continuously changing forcings.

A modification was made in the climate model described in Section 2 to allow its coupling with the diffusive deep-ocean model. This change concerns the mixed-layer depth below sea ice which was increased from 30 to 150 m. Such a change has a non-negligible impact on the sea-ice simulation for present-day conditions. In order to recover a realistic simulation, values of 0.70 and 0.45 were assigned to the upper and lower limits of the ice albedo (instead of 0.75 and 0.40).

Since the climate system acts in long-term response to external changes, the post-1700 climate response depends to some degree on the forcing history prior to the year 1700. Sensitivity tests have shown that this "history effect" is a vital consideration during the first few decades, but becomes unimportant after 30-40 years (see also WIGLEY & RAPER, 1990; FICHEFET & TRICOT, 1992). In view of this, we have decided to analyse only the model outputs posterior to the year 1765. It should be noted that the results presented in the following are further discussed in SMITS et al. (1993).

5.1   Response to the solar and astronomical forcings

A first experiment was performed to quantify the influence of the solar irradiance changes on the time evolution of climate at the secular time scale. In this experiment, the earth's orbital elements were kept fixed to their 1700 values and the CO_2 concentration was taken equal to 280 ppmv, i.e., its pre-industrial value (HOUGHTON et al., 1990). WILLSON & HUDSON (1988) correlated, over the time period 1980-1986, the ACRIM (Active Cavity Radiometer Irradiance Monitor) daily mean data of total solar irradiance with the Wolf number, R_z, and suggested the following relationship:

$$S_o = 1366.82 + 7.71 \ 10^{-3} \times R_z$$

where S_o is in Wm^{-2}. This parameterization was used here and assumed to be valid over the last three centuries. The Wolf number data were supplied by the Sunspot Index Data Centre of Uccle, Belgium (A. Koeckelenbergh and P. Cugnon, 1992, pers. comm.).

The lower curve in Fig. 5 shows the reconstructed solar irradiance between 1765 and 1990. The maximum variation over this time period is about 1.5 Wm^{-2} ($\cong 0.1\%$) and occurs during solar cycle 19 (around the year 1957). The upper curve in the same figure gives the time evolution of the hemispheric annual mean surface temperature predicted by the model. A series of 11-year peaks is observed, with the largest amplitude ($\sim 0.05°C$) taking place around the year 1957. The maximum temperature change between the low irradiance period in the early 1800s and the higher values in the 1950s is about 0.07°C. Note that, over the last century, there is a gradual warming of about 0.03°C. A detailed examination of Fig. 5 reveals that the surface temperature follows the solar activity cycle with a small time delay. On the other hand, it can be seen that the temperature variations are considerably smoothed relative to the forcing. These two features are consequences of the large thermal inertia of the ocean. In order to isolate a possible long-term temperature change in our simulation, we have computed the difference between the mean temperature of the time interval 1876-1990 and that of the time period 1765-1875. This difference amounts to +0.0052°C and is therefore indicative of a weak warming trend. Figure 6a displays the time evolution of the annual mean total sea-ice area simulated by the model. The ice area is maximum between 1805 and 1825, when the irradiance is rather weak. It is minimal around the year 1957, when the solar activity is at its maximum. Between the two time intervals chosen for averaging, the ice extent decreases by about 3,600 km^2 ($\cong 0.03\%$). The continental snow area (Fig. 6b) shows approximately the same behaviour.

A second experiment was carried out in which the solar irradiance and the CO_2 concentration were kept constant to their 1700 values, while the earth's orbital elements were permitted to vary according to BERGER (1978). The changes in these elements generate a decrease in the hemispheric annual mean insolation received at the top of the atmosphere of only 0.0007 Wm^{-2} between 1765 and 1990. Figure 7 demonstrates that the behaviour can be quite different for a particular month. For example, the September insolation decreases by about 1 Wm^{-2} over the time period considered, whereas the April one increases by approximately the same amount. These insolation variations induce in the model a hemispheric annual mean surface cooling of 0.0032°C between the two time periods 1765-1875 and 1876-1990. The associated increase in sea-ice area is about 0.2%. This large (compared to the previous experiment) change in ice extent is related to the long-term decrease in insolation during the melting period. An additional experiment was made by taking into account the time variations in both the solar irradiance and the earth's orbital elements. The long-term response of the hemispheric annual mean surface temperature (+0.0017°C) simulated in this case is almost equal to the sum of the responses obtained in the first two experiments (+0.0020°C).

5.2 Response to the solar, astronomical, and greenhouse-gas forcings

We now consider the solar and astronomical effects superimposed on the forcing due to the gradual increase in greenhouse-gas concentrations. This forcing was expressed in terms of changes in equivalent CO_2 concentration. (The equivalent CO_2 concentration is the CO_2 concentration that would have the same radiative effect at the tropopause than the one caused by all the greenhouse gases, except water vapour which is an internal variable of the model). The time evolution of the equivalent CO_2 concentration between 1700 and 1990 was determined using concentration histories and concentration-forcing relationships for the most important greenhouse gases (i.e., CO_2, CH_4, N_2O, stratospheric H_2O, and CFCs) recommended by HOUGHTON et al. (1990). The resulting radiative forcing at the tropopause is depicted in Fig. 8. One notes an increase of about 2.5 Wm^{-2} between 1765 and 1990. Two experiments were conducted with the model submitted to this forcing. In the first one, the solar irradiance and the earth's orbital elements were kept constant to their 1700 values, while in the second one they were allowed to vary.

Figure 9 illustrates the transient response of the hemispheric annual mean surface temperature simulated by the model with and without solar and astronomical effects. One sees that the solar and astronomical responses show up only as minor perturbations. The surface warming produced by the model between 1765 and 1990 in the experiment with solar and astronomical effects incorporated amounts to about 0.9°C. Over the time interval 1890-1990, the temperature increase is about 0.7°C. This warming appears somewhat overestimated compared to the observed global temperature trend (0.45±0.15°C; HOUGHTON et al., 1992). Possible reasons for this discrepancy are given in the concluding remarks. The transient response of the zonally averaged annual mean surface temperature is given in Fig. 10. This figure shows a continuous warming over the whole hemisphere that is particularly pronounced at high latitudes because of the sea-ice and snow albedo/temperature feedback. The magnitude of the warming reaches nearly 2°C in 1990 around 80°N. At mid-latitudes, the surface temperature increase is about 1°C in 1990 and, at lower latitudes, it ranges

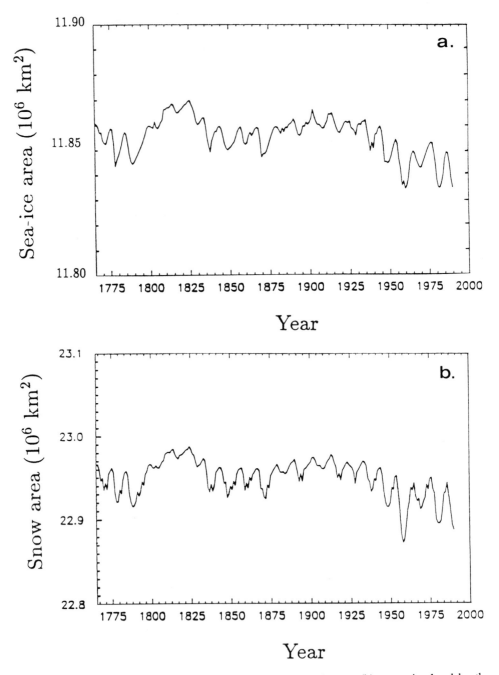

Fig. 6 Time evolution of the annual mean total sea-ice (a) and snow (b) areas simulated by the model in response to the solar irradiance changes displayed in Fig. 5

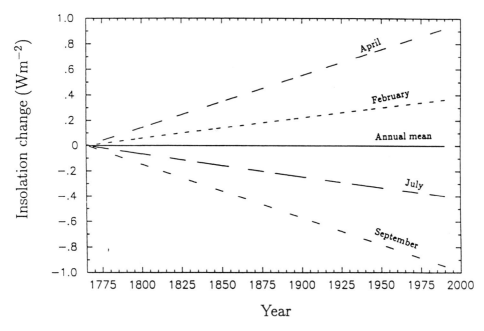

Fig. 7 Variations in the hemispheric mean insolation received at the top of the atmosphere since 1765 due to the changes in the earth's orbital elements (computed after BERGER, 1978)

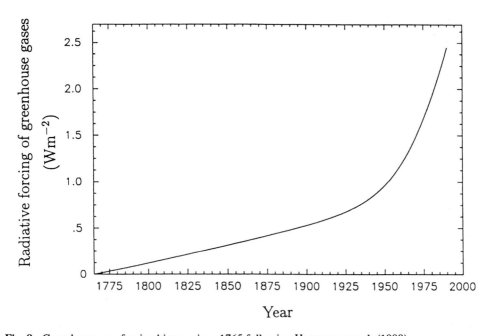

Fig. 8 Greenhouse-gas forcing history since 1765 following HOUGHTON et al. (1990)

between 0.7 and 0.8°C. Figure 11 gives the time evolution of the annual mean total sea-ice and snow areas predicted by the model. It reveals that, over the time period 1765-1990, the ice and snow extents decrease by about 7%.

The results of these two experiments indicate that a large part of the warming simulated by the model between 1765 and 1990 is induced by the build-up of greenhouse gases, the relative part of the solar and astronomical forcings being extremely small.

6. Conclusions

A two-dimensional zonally averaged climate model has been used to determine the possible relative importance of the solar and astronomical forcings on the climate of the last three centuries. It has been shown, in accordance with previous modelling studies (e.g., WIGLEY & RAPER, 1990; GÉRARD & HAUGLUSTAINE, 1991), that the influence of the solar irradiance changes on the time evolution of climate at the secular time scale is likely to be extremely weak, unless some other mechanism is operating beyond that leading to the photospheric effects encapsulated in the WILLSON & HUDSON (1988) parameterization. As noted by several authors (e.g., EDDY, 1976; GILLILAND, 1982; REID, 1987, 1991; BALIUNAS & JASTROW, 1990; WIGLEY & KELLY, 1990; FRIIS-CHRISTENSEN & LASSEN, 1991) additional low frequency variations are possible given the absence of reliable irradiance data prior to the satellite era. Recently, using knowledge of the mechanisms of irradiance variations gleaned from extant solar data, LEAN et al. (1992) estimated that the sun's radiative output during the Maunder Minimum was about 0.25% below its present-day mean value. A forcing change of this magnitude induces in our model a decrease in the equilibrium surface temperature of 0.4°C. The question of solar effects on climate therefore remains open. Irradiance measurements over many decades are required to finally resolve this issue. With regard to the astronomical forcing, the experiments performed here suggest that it has a negligible impact on the climate of the last three centuries.

When the solar and astronomical effects are combined with the forcing due to the build-up of greenhouse gases, the model simulates a hemispheric annual mean surface warming of 0.7°C between 1890 and 1990. Over the same time interval, the observations shows a global temperature increase of 0.45±0.15°C (HOUGHTON et al., 1992). Part of the discrepancy between the model and data could be due to an underestimation of the lag in the model response to external forcing changes. Indeed, our model has an e-folding time for an instantaneous CO_2 doubling of 13 years, which is in the lower range of current estimates (10-100 years; SCHLESINGER, 1989). We will investigate this important point in the future by replacing the simple diffusive deep-ocean model employed here by a zonally averaged three-basin ocean circulation model (FICHEFET & HOVINE, 1993). This modification will also allow for the internal variability of climate related to ocean dynamics and ocean-atmosphere interactions to be accounted for. It is worth pointing out that the absence of internal

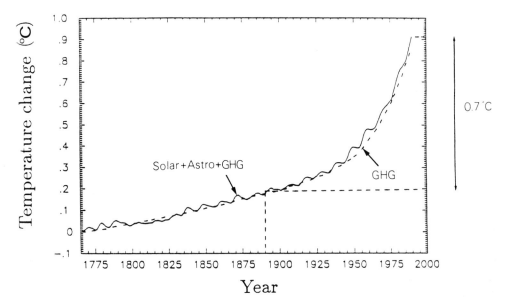

Fig. 9 Transient response of the hemispheric annual mean surface temperature to the greenhouse-gas forcing alone (dashed curve) and to the solar, astronomical, and greenhouse-gas forcings combined (solid curve)

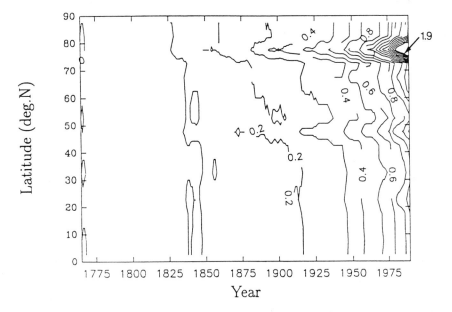

Fig. 10 Transient response of the zonally averaged annual mean surface temperature to the solar, astronomical, and greenhouse-gas forcings combined. Units are °C

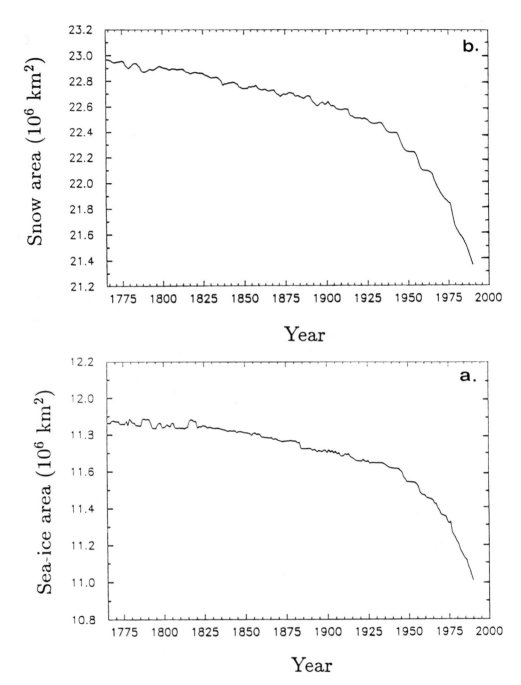

Fig. 11 Time evolution of the annual mean total sea-ice (a) and snow (b) areas simulated by the model in response to the solar, astronomical, and greenhouse-gas forcings taken together

variability in the present version of the model could also be partly responsible for the disagreement mentioned above. Finally, we stress the fact that, in the present work, we have not considered all the forcings that can influence the climate system at the secular time scale. The concentration of stratospheric aerosols can be greatly enhanced over large areas for a few years following large explosive volcanic eruptions and can significantly affect the earth's radiative balance (e.g., HANSEN & LACIS, 1990; SHINE et al., 1990). In addition, it has long been recognized that tropospheric aerosols may exert a global cooling influence on climate, e.g., because of their scattering of short-wave radiation and the resultant increase in planetary albedo. Since 1850, industrial activities, especially emissions of SO_2, contributed to substantial local increases in the amount of tropospheric aerosols. However, the quantitative impact of these increases on the global radiative budget of the earth is yet difficult to assess because the anthropogenic aerosols are distributed quite non-uniformly over the earth and are relatively short-lived (e.g., CHARLSON et al., 1992).

Acknowledgements

We thank A. Koeckelenbergh and P. Cugnon for providing us with the Wolf number data. This research was partly supported by the Climate Programme of the Commission of the European Communities under contract EPOC-0003-C(MB). Th. Fichefet is sponsored by the National Fund for Scientific Research (Belgium). The National Fund for Scientific Research (Belgium) and the "Fonds de Développement Scientifique" of the "Université Catholique de Louvain" (Louvain-la-Neuve) supplied the computer time needed to complete the calculations. The graphics were made with the NCAR package.

References

BALIUNAS, S. & JASTROW, R. (1990): Evidence for long-term brightness changes of solar-type stars. Nature 348, 520-522

BERGER, A. (1978): Long-term variations of daily insolation and Quaternary climatic changes. J. Atmosph. Sci. 35, 2362-2367

BERGER, A. (1988): Milankovitch theory and climate. Rev. Geophys. 26, 624-657

BRIEGLEB, B. & RAMANATHAN, V. (1982): Spectral and diurnal variations in clear sky planetary albedo. J. Appl. Meteor. 21, 1160-1171

BROECKER, W. S. (1991): The great ocean conveyor. Oceanography 4, 79-89

CHARLSON, R. J.; SCHWARTZ, S. E.; HALES, J. M.; CESS, R. D.; COAKLEY, J. A.; HANSEN, J. E. & HOFMANN, D. J. (1992): Climate forcing by anthropogenic aerosols. Science 255, 423-430

DANARD, M.; GRAY, M. & LYV, G. (1984): A model for predicting ice accretion and ablation in water bodies. Monthly Weather Review 112, 1160-1169

EDDY, J. A. (1976): The Maunder Minimum. Science 192, 1189-1202

FICHEFET, T. & HOVINE, S. (1993): The glacial ocean: a study with a zonally averaged three-basin ocean circulation model. In: Peltier, W. R. (ed.): Ice in the climate system. NATO ASI Series, Vol. I/12. Springer Verlag, Berlin, Heidelberg, 433-458

FICHEFET, T. & TRICOT, C. (1992): Influence of the starting date of model integration on projections of greenhouse-gas-induced climatic change. Geophys. Res. Lett. 19, 1771-1774

FOUKAL, P. & LEAN, J. (1990): An empirical model of total solar irradiance variation between 1874 and 1988. Science 242, 556-558

FRIIS-CHRISTENSEN, E. & LASSEN, K. (1991): Length of the solar cycle: an indicator of solar activity closely associated with climate. Science 254, 698-700

GALLÉE, H.; VAN YPERSELE, J. P.; FICHEFET, T.; TRICOT, C. & BERGER, A. (1991): Simulation of the last glacial cycle by a coupled, sectorially averaged climate-ice sheet model. I. The climate model. J. Geophys. Res. 96, 13, 139-13, 161

GASPAR, P. (1988): Modelling the seasonal cycle of the upper ocean. J. Phys. Oceanogr. 18, 161-180

GÉRARD, J. C. & HAUGLUSTAINE, D. A. (1991): Transient climate response to solar irradiance: reconstruction for the last 120 years. Climate Research 1, 161-167

GILLILAND, R. L. (1982): Solar, volcanic and CO_2 forcing of recent climatic changes. Climatic Change 4, 111-131

HANSEN, J. E. & LACIS, A. A. (1990): Sun and dust *versus* greenhouse gases: an assessment of their relative roles in global climate change. Nature 346, 713-719

HANSEN, J.; FUNG, I.; LACIS, A.; RIND, D; LEBEDEFF, S.; RUEDY, R.; RUSSELL, G. & STONE, P. (1988): Global climate changes as forecast by Goddard Institute for Space Studies three dimensional model. J. Geophys. Res. 93, 9341-9364

HANSEN, J.; LACIS, A.; RIND, D; RUSSELL, G.; STONE, P.; FUNG, I.; RUEDY, R. & LERNER, J. (1984): Climate sensitivity: analysis of feedback mechanisms. In: Hansen, J. E. & Takahashi, T. (eds.): Climate processes and climate sensitivity. Geophys. Monogr. Series 29. Am. Geophys. Union (AGU), Washington, D.C., 130-163

HARVEY, L. D. D. (1988a): A semi-analytic energy balance climate model with explicit sea ice and snow physics. J. Climate 1, 1065-1085

HARVEY, L. D. D. (1988b): On the role of high latitude ice, snow, and vegetation feedbacks in the climatic response to external forcing changes. Climatic Change 13, 191-224

HOUGHTON, J. T.; CALLENDER, B. A. & VARNEY, S. K. (eds.) (1992): Climate change 1992. The supplement report to the IPCC scientific assessment. Cambridge Univ. Press, Cambridge, 200 p.

HOUGHTON, J. T.; JENKINS, G. T. & EPHRAUMS, J. J. (eds.) (1990): Climate change. The IPCC scientific assessment. Cambridge Univ. Press, Cambridge, 364 p.

LEAN, J. (1991): Variations in the sun's radiative output. Rev. Geophys. 29, 505-535

LEAN, J.; SKUMANICH, A. & WHITE, O. (1992): Estimating the sun's radiative output during the Maunder minimum. Geophys. Res. Lett. 19, 1591-1594

LEDLEY, T. S. (1985): Sensitivity of a thermodynamic sea ice model with leads to time step size. J. Geophys. Res. 90, 2251-2260

LEVITUS, S. (1982): Climatological Atlas of the World Ocean. NOAA Prof. Pap. 13. U.S. Govt. Printing Office, Washington, D.C., 173 p.

MANABE, S. (1969): The atmospheric circulation and the hydrology of the earth's surface. Mon. Wea. Rev. 97, 739-774

MANABE, S.; STOUFFER, R. J.; SPELMAN, M. J. & BRYAN, K. (1991): Transient responses of a coupled ocean-atmosphere model to gradual changes of atmospheric CO_2. Part I: Annual mean response. J. Climate 4, 785-818

MAROTZKE, J. & WILLEBRAND, J. (1991): Multiple equilibria of the global thermohaline circulation. J. Phys. Oceanogr. 21, 1372-1385

MIKOLAJEWICZ, U. & MAIER-REIMER, E. (1990): Internal secular variability in an ocean general circulation model. Climate Dynamics 4, 145-156

MORCRETTE, J. J. (1984): Sur la paramétrisation du rayonnement dans les modèles de la circulation générale atmosphèrique, Thèse de Doctorat d'Etat. Université des Sciences et des Techniques de Lille, Lille, 373 p.

MYSAK, L. A.; STOCKER, T. F. & HUANG, F. (1993): Century-scale variability in a randomly forced, two-dimensional thermohaline ocean circulation model. Climate Dynamics 8, 103-116

NEEMAN, B. U.; OHRING, G. & JOSEPH, J. H. (1988): The Milankovitch theory and climate sensitivity. I. Equilibrium climate model solution for the present surface conditions. J. Geophys. Res. 93, 11153-11174

NEWMAN, M. J. & ROOD, R. T. (1977): Implications of solar evolution for earth's early atmosphere. Science 198, 1035-1037

OERLEMANS, J. (1982): Response of the Antarctic ice sheet to a climate warming: a model study. J. Climatol. 2, 1-11

OHRING, G. & ADLER, S. (1978): Some experiments with azonally-averaged climate model. J. Atmosph. Sci. 35, 186-205

OORT, A. H. (1983): Global atmospheric circulation statistics 1958-1973, NOAA Prof. Pap. 14, Washington, D.C., 180 p.

PARKINSON, C. L. & WASHINGTON, W. M. (1979): A large-scale numerical model of sea ice. J. Geophys. Res. 84, 311-337

PENG, L.; CHOU, M. D. & ARKING, A. (1982): Climate studies with a multi-layer energy balance model. Part I: Model description and sensitivity to the solar constant. J. Atmosph. Sci. 39, 2639-2656

PENG, L.; CHOU, M. D. & ARKING, A. (1987): Climate warming due to increasing atmospheric CO_2 simulations with a multi-layer coupled atmosphere-ocean seasonal energy balance model. J. Geophys. Res. 92, 5505-5521

POTTER, G. L. & CESS, R. D. (1984): Background tropospheric aerosols: incorporation within a statistical-dynamical climate model. J. Geophys. Res. 89, 9521-9526

REID, G. C. (1987): Influence of solar variability on global sea surface temperatures. Nature 329, 142-143

REID, G. C. (1991): Solar total irradiance variations and the global sea surface temperature record. J. Geophys. Res. 96, 2835-2844

ROBOCK, A. (1980): The seasonal cycle of snow cover, sea ice and surface albedo. Mon. Wea. Rev. 108, 267-285

SALTZMAN, B. (1980): Parameterization of the vertical flux of latent heat at the earth's surface for use in statistical-dynamical climate models. Arch. Meteor. Geophys. Bioklim. A29, 41-53

SALTZMAN, B. & ASHE, S. (1976): Parameterization of the monthly mean vertical heat transfer at earth's surface. Tellus 28, 323-331

SCHATTEN, K. H. (1988): A model for solar constant changes. Geophys. Res. Lett. 15, 121-124

SCHLESINGER, M. E. (1989): Model projections of the climatic changes induced by increasing atmospheric CO_2. In: Berger, A.; Schneider, S. & Duplessy, J. C. (eds.):

Climate and Geosciences. NATO-ASI Series 285, Kluwer Acad. Pub., Dordrecht, 375-415

SELA, J. & WIIN-NIELSEN, A. (1971): Simulation of the atmospheric annual energy cycle. Mon. Wea. Rev. 99, 460-468

SELLERS, W. D. (1973): A new global climate model. J. Appl. Meteorol. 22, 1557-1574

SEMTNER, A. J. (1976): A model for the thermodynamic growth of sea ice in numerical investigations of climate. J. Phys. Oceanogr. 6, 379-389

SHINE, K.; DERWENT, R. G.; WUEBBLES, D. J. & MORCRETTE, J. J. (1990): Radiative forcing of climate. In: Houghton, J. T.; Jenkins, G. J. & Ephraums, J. J. (eds.): Climate change. The IPCC scientific assessment. Cambridge Univ. Press, Cambridge, 45-74

SMITS, I.; FICHEFET, T.; TRICOT, C. & VAN YPERSELE, J.P. (1993): A model study of the time evolution of climate at the secular time scale. Atmosphera 6, 255-272

STOCKER, T. F. & WRIGHT, D. G. (1991): Rapid transitions of the ocean's deep circulation induced by changes in surface water fluxes. Nature 251, 729-732

TAYLOR, K. (1976): The influence of subsurface energy storage on seasonal temperature variations. J. Appl. Meteorol. 15, 1129-1138

TRICOT, C. (1989): The transient response of climate to greenhouse gas concentration changes: a preliminary study with a two-dimensional coupled atmosphere-ocean model. In: Crutzen, P. J.; Gérard, J. C. & Zander, R. (eds.): Our changing atmosphere. Proceedings of the 28th Liège International Colloquium. Université de Liège, Liège, 333-338

TRICOT, C. & BERGER, A. (1988) Sensitivity of present-day climate to astronomical forcing. In: Wanner, H. & Siegenthaler, U. (eds.): Long and short-term variability of climate. Lect. Notes Earth Sci. 16. Springer-Verlag, Berlin, 132-152

WHITE, A. A. & GREEN, J. S. A. (1984): Transfer coefficient eddy flux parameterizations in a simple model of the zonal average atmospheric circulation. Quart. J. Roy. Meteor. Soc. 110, 1035-1052

WIGLEY, T. M. L. & KELLY, P. M. (1990): Holocene climatic change, ^{14}C wiggles and variations in solar irradiance. Phil. Trans. Roy. Soc. London A330, 547-560

WIGLEY, T. M. L. & RAPER, S. C. B. (1990): Climatic change due to solar irradiance changes. Geophys. Res. Lett. 17, 2169-2172

WILLSON, R. C. & HUDSON, H. S. (1988): Solar luminosity variations in solar cycle 21. Nature 332, 810-812

WILLSON, R. C. & HUDSON, H. S. (1991): The sun's luminosity over a complete solar cycle. Nature 351, 42-44

WILLSON, R. C.; GULKIS, S.; JANSSEN, M.; HUDSON, H. S. & CHAPMAN, G. A. (1981): Observations of solar irradiance variability. Science 211, 700-702

WILSON, C. A. & MITCHELL, J. F. B. (1987): A doubled CO_2 climate sensitivity experiment with a global climate model including a simple ocean. Journal of Geophysical Research 92, 13315-13343

WETHERALD, R. T. & MANABE, S. (1975): The effects of changing the solar constant on the climate of a general circulation model. J. Atmosph. Sci. 32, 2044-2059

Author's address:

Dr. T. Fichefet, Université Catholique de Louvain, Institut d'Astronomie et de Géophysique G. Lemaître, chemin du Cyclotron 2, B-1348 Louvain-la-Neuve

PERIODICAL TITLE ABBREVIATIONS

Ann. Geophys.	• Annales de Géophysique
Ann. Glaciol.	• Annals of Glaciology
Ann. Rev. Astron. Astrophys.	• Annual Review of Astronomy and Astrophysics
Appl. Geochem.	• Applied Geochemistry
Arch. Meteor. Geophys. Bioklim.	• Archiv für Meteorologie, Geophysik und Bioklimatologie
Astronaut. Aeron.	• Astronautics and Aeronautics
Astron. Astrophys.	• Astronomy and Astrophysics
Astrophys. J.	• Astrophysical Journal
Astrophys. Lett.	• Astrophysical Letters
Aust. J. Biol. Sci.	• Australian Journal of Biological Sciences
Biul. Geol.	• Biuletyn Geologiczny
CNR	• Consiglio Nazionale delle Ricerche
COSPAR	• Committee on Space Research
C. R. Acad. Sci.	• Comptes Rendus de l'Académie des Sciences
Earth Planet. Sci. Lett.	• Earth and Planetary Science Letters
Earth Surf. Proc. Landf.	• Earth Surface Processes and Landforms
Ed. Sp. Acad. Roy. Serbe	• Edition Spéciale de l'Académie Royale Serbe
ESA	• European Space Agency
Forstwiss. Cbl.	• Forstwissenschaftliches Centralblatt
GAFD	• Geophysical Astrophysical Fluid Dynamics
Geochim. Cosmochim. Acta	• Geochimica et Cosmochimica Acta
Geol. Jahrb.	• Geologisches Jahrbuch
Geol. Rdsch.	• Geologische Rundschau
Geophys. Monogr. Series	• Geophysical Monograph Series
Geophys. Res. Lett.	• Geophysical Research Letters
IAEA	• International Atomic Energy Agency (Vienna)
IASH	• International Association of Scientific Hydrology
ICTIMA	• Istituto di Chimica e Tecnologie Inorganiche dei Materiali Avanzati
IPCC	• Intergovernmental Panel on Climate Change
IUGG	• International Union of Geophysics and Geodesy
J. Appl. Meteor.	• Journal of Applied Meteorology
J. Archaeol. Sci.	• Journal of Archaeological Science
J. Atm. Terr. Electr. (Phys.)	• Journal of Atmospheric and Terrestrial Electricity (Physics)

J. Atmosph. Sci.	• Journal of the Atmospheric Sciences
J. Clim. Appl. Meteor.	• Journal of Climate and Applied Meteorology
J. Climate	• Journal of Climate
J. Climatol.	• Journal of Climatology
J. Comput. Phys.	• Journal of Computational Physics
J. Geomagn. Geoelectr.	• Journal of Geomagnetism and Geoelectricity
J. Geophys. Res.	• Journal of Geophysical Research
J. Glaciol.	• Journal of Glaciology
J. Meteor. Soc.	• Journal of the Meteorological Society
J. Phys. Oceanogr.	• Journal of Physical Oceanography
Lect. Notes Earth Sci.	• Lecture Notes in Earth Sciences
Medd. Grønl.	• Meddelelser om Grønland
Mem. Soc. Geogr. It.	• Memorie Società Geografica Italiana
NASA Conf. Pub.	• National Aeronautics and Space Administration Conference Publication
NATO ASI	• North Atlantic Treaty Organization Advanced Study Institute
NOAA	• National Oceanic and Atmospheric Administration
Nucl. Sci. Abstr.	• Nuclear Science Abstracts
Publ. Astron. Soc. Pacific	• Publications of the Astronomical Society of the Pacific
Phil. Trans. Roy. Soc. London	• Philosophical Transactions of the Royal Society of London
PMOD/WRC	• Physikalisch-Meteorologisches Observatorium Davos/World Radiation Center
Proc. Int. Symp. LIA	• Proceedings of the International Symposium on the Little Ice Age
Proc. Nat. Acad. Sci. USA	• Proceedings of the National Academy of Sciences of the USA
Prof. Pap.	• Professional Papers
Quart. J. Roy. Meteor. Soc.	• Quarterly Journal of the Royal Meteorological Society
Quat. Res.	• Quaternary Research
Quat. Sci. Rev.	• Quaternary Science Reviews
Rev. Geophys.	• Reviews of Geophysics
Sci. Am.	• Scientific American
Sol. Phys.	• Solar Physics
Space Sci. Rev.	• Space Science Reviews
Stud. Geophys. Geod.	• Studia Geophysica et Geodaetica
Theor. Appl. Climatol.	• Theoretical and Applied Climatology
Trans. Inst. Brit. Geogr.	• Transactions of the Institute of British Geographers

Herausgegeben von der Akademie der Wissenschaften und der Literatur, Mainz.
Mathematisch-naturwissenschaftl. Klasse, Prof. Dr. Dr. Burkhard Frenzel, Stuttgart.

PALÄOKLIMAFORSCHUNG

Preisänderungen vorbehalten

PALÄOKLIMAFORSCHUNG

Herausgegeben von der Akademie der Wissenschaften und der Literatur, Mainz.
Mathematisch-naturwissenschaftl. Klasse, Prof. Dr. Dr. Burkhard Frenzel, Stuttgart.

Band 8: Evaluation of land surfaces cleared from forests by prehistoric man in Early Neolithic times and the time of migrating Germanic tribes

Edited by Prof. Dr. Dr. B. Frenzel, co-edited by Prof. Dr. L. Reisch and Dr. B. Gläser.
1992. XII, 225 S., 57 Abb., 5 Tab., kt. DM 74,–

Band 9: Oscillations of the Alpine and Polar Tree Limits in the Holocene

Edited by Prof. Dr. Dr. B. Frenzel, co-edited by Dr. M. Eronen, Prof. Dr. K.-D. Vorren and Dr. B. Gläser.
1993. XII, 234 S., 81 Abb., 8 Tab., kt. DM 78,–

Band 10: Evaluation of land surfaces cleared from forests in the Mediterranean region during the time of the Roman empire

Edited by Prof. Dr. Dr. B. Frenzel, co-edited by Prof. Dr. L. Reisch and
Dipl. Agr.-Biol. M. M. Weiß.
1994. X, 170 S., 48 Abb., 2 Tab., kt. DM 78,–

Band 11: Solifluction and climatic variation in the Holocene

Edited by Prof. Dr. Dr. B. Frenzel, co-edited by Dr. J. A. Matthews and Dr. B. Gläser.
1993. X, 387 S., 107 Abb., 38 Fotos, 35 Tab., kt. DM 118,–

Band 12: Evaluation of land surfaces cleared from forests in the Roman Iron Age and the time of migrating Germanic tribes based on regional pollen diagrams

Edited by Prof. Dr. Dr. B. Frenzel, co-edited by Dr. S. T. Andersen,
Prof. Dr. B. E. Berglund and Dr. B. Gläser.
1994. VIII, 134 S., 27 Abb.., 1 Tab. kt. DM 58,–

Band 13: Climatic trends and anomalies in Europe 1675 – 1715

Edited by Prof. Dr. Dr. B. Frenzel, co-edited by Dr. B. Gläser and Prof. Dr. C. Pfister.
1994. XII, 479 S., 161 Abb., 55 Tab., kt. DM 148,–

Band 14: European river activity and climatic change during the Lateglacial and early Holocene

Edited by Prof. Dr. Dr. B. Frenzel, co-edited by Prof. Dr. J. Vandenberghe,
Dr. K. Kasse, Dr. S. Bohncke and Dr. B. Gläser.
1995. VII, 226 S., 80 Abb., 11 Tab., 8 Fotos,
kt. DM 78,–

GUSTAV FISCHER

SEMPER BONIS ARTIBUS

Preisänderungen vorbehalten.